Fertility Management in Dairy Cattle

Fertility Management in Dairy Cattle

R.J. Esslemont
University of Reading

J.H. Bailie
Coopers Animal Health

M.J. Cooper
Veterinary Surgeon, Staffordshire

COLLINS
8 Grafton Street, London W1

Collins Professional and Technical Books
William Collins Sons & Co. Ltd
8 Grafton Street, London W1X 3LA

First published in Great Britain by
Collins Professional and Technical Books 1985

Distributed in the United States of America
by Sheridan House, Inc.

British Library Cataloguing in Publication Data
Esslemont, R.J.
Fertility management in dairy cattle.
1. Dairy cattle—Breeding.   2. Fertility
I. Title   II. Bailie, J.H.
III. Cooper, M.J.
636.2'142   SF201

ISBN 0-00-383032-2

Typeset by Columns of Reading
Printed and bound in Great Britain by
Mackays of Chatham, Kent

Contents

Preface

When veterinary surgeons meet farmers in carrying out their role as cattle advisers they are often asked for help in deciding whether cattle should be kept or culled. If the vet finds weaknesses in cattle fertility in terms of heat detection or pregnancy rate he should be able to turn to economic principles before he advises a farmer on the best way of improving the situation. With quotas operating in U.K. dairying it has become even more important to run the dairy herd with the utmost efficiency. This means that the cows must calve at the right time of the year; must be seen in heat; must be efficiently served and got in calf as quickly as practicable. Culling must be carried out on a sensible, economic basis. This book attempts to show the steps in that process using modern aids and up to date knowledge.

In chapter 1 the present day economics of dairying are described and the circumstances of the U.K. dairy industry outlined. This sets the scene for chapter 2 where the economics of fertility are listed in detail so that advisers can give sensible advice to dairy farmers about the cost effectiveness of various alternatives. In chapter 3 the physiology of the reproductive cycle in the cow is described and illustrated, allowing chapter 4 to develop the use of this knowledge and incorporate it in planned cattle breeding using modern aids such as prostaglandins. Chapter 5 covers all aspects of the husbandry of cattle fertility, looking at the policy of the farmer, heat detection, pregnancy rate and culling strategy. This is a major area in the book, and covers a wide field. Chapter 6 looks at the work of the veterinarian in cattle management today, dealing with the cows he should see on his regular visits and some of the treatments to give cattle with problems. Chapter 7 moves on to deal with recording systems for fertility information on the farm, showing what information to record for which cow, when to record it and how to analyse the results. The alarm lists to produce for regular management of cattle are highlighted, as are the analyses of conception rate and heat detection rate. Modern microcomputer systems are described and their virtues and vices outlined. These cost effective tools are available widely to help the cattle adviser improve fertility management. In chapter 8 a case study is described which shows how improved husbandry and management has led to large improvements in profitability. This case study may be useful for advisers carrying out similar analysis of herds they are working with.

Acknowledgements

Many workers in fertility management in cattle have made contributions to this book through their own research work including Roger Eddy, Colin Gould, Roger Blowey and Bridget Drew who have provided facts and ideas.

We are also very grateful to Liz Wild, Sheila Wagstaff, Wendy Gibbons and Linda Grimbeby for their help with collation, typing and proof-reading, and Jean Taylor for her excellent artwork.

1 Present day dairy economics

Trends in dairying

The U.K. dairy industry always seems to be at a critical 'crossroads'. While it is only able to produce 88% of the dairy requirements in the U.K. it is currently, with the other countries of the E.E.C., providing 120% of European milk requirements. To try to control the surpluses, a quota system is now being applied. On average in the U.K., 1.6 tonnes of expensive concentrate feed are used per cow per year. While some will remain necessary to produce economic amounts of milk, there is considerable scope for rationalisation of milk output by the more efficient use of cheap home-grown forages.

Whether considered in terms of yield per cow or in terms of herd size, U.K. dairying appears to be at the top end of the E.E.C. league table (table 1.1), but it is in the middle of the list in terms of milk sales less concentrate costs expressed per hectare (table 1.2). The main reason for this is the low stocking rate in Britain,

Table 1.1 Dairy statistics of major E.E.C. dairying nations

Country	Herd size (cows)	Yield per cow (kg)	Production of cows: milk (million tonnes)
Germany	13.9	4610	25.8
France	16.0	3776	32.9
Netherlands	39.4	5230	10.8
U.K.	55.5	5099	18.5
Ireland	17.0	3547	5.7
Denmark	26.7	5160	5.4

Source: E.E.C. Dairy Facts and Figures 1983

Table 1.2 Margin over concentrates (M.O.C.) at constant prices for seven areas in the E.E.C.

	Stocking rate (cows/hectare)	M.O.C. (£/cow)	M.O.C. (£/hectare)
England and Wales	1.95	538	1054
Schleswig-Holstein	2.28	406	926
Bavaria	2.13	503	1076
Brittany	1.94	542	1051
Netherlands	2.87	579	1662
Ireland	1.75	444	781
Denmark	3.24	427	1383

Source: A Comparison of European Dairying (1983), F.M.S. Information Unit, M.M.B.

Table 1.3 Energy utilisation from forage in seven areas in the E.E.C.

	Energy utilised Gigajoules/hectare (GJ/ha)
England and Wales	65.8
Schleswig-Holstein	56.9
Bavaria	77.0
Brittany	74.5
Netherlands	102.1
Ireland	64.3
Denmark	91.6

Source: M.M.B.

compared particularly to Holland and Denmark, reflecting the poor use of forage in the U.K. (table 1.3).

Under pressure from reduced returns this present picture of the wasteful use of farm resources may change quickly, and we may see a rapid reduction in concentrate consumption and a more efficient use of such inputs as fertilizer to produce more energy per hectare.

Yields per cow have been increasing in the U.K. at two to three per cent per year. If this continues then by 1992 the U.K. should be able to produce the same amount of milk as in 1982 but from 680000 fewer cows (table 1.4). This would lead to 440000 fewer calves being available for beef production (65% of dairy calves go into beef) and to 160000 fewer cull cows being available for that trade. At present the quota system is designed to stabilise production at the 1981 level without allowing the farmer to make up for yield quotas by increasing milk quality. However, the quality constituents are used as one of the main bases for calculating payments per litre, with particular emphasis on fat and protein (table 1.5).

Table 1.4 Changes in cow numbers required to maintain milk output assuming yield per cow rises at 2–3% per year

Region	Dairy cow numbers 1982	(millions) 1992	Fall in numbers (millions)
Total E.E.C.	25.39	20.23	5.16
U.K.	3.34	2.66	0.68

Table 1.5 Pricing of milk 1985

Component	Price per % point (pence per litre)
Fat	1.889
Protein	1.849
Lactose	0.281

The U.K. dairy industry has changed markedly in terms of mechanisation over the last twenty years. It is now much better equipped mechanically, with the majority of cows milked through herringbone parlours while the number of labour intensive cowshed systems has diminished dramatically (table 1.6).

There were only 5.6% of herds with more than 100 cows in 1972, but now there are 14% of such herds (table 1.7). The average herd size is nearly 70 cows in England and Wales with yield per cow being highest in the larger herds. The yield in smaller herds is

Table 1.6 Changes in distribution of herd parlour types 1972–81

| | % of herds | | % of cows | |
	1972	1981	1972	1981
Cowshed	66	24	42	12
Herringbone	10	34	22	52
Abreast	18	25	26	22
Other	6	17	10	14

Source: Milk Production 1980–81 H.M.S.O.

Table 1.7 Distribution of herds and cows by herd size (England and Wales)

| | % herds | | | % cows | | |
Herd size	1972	1976	1980	1972	1976	1980
10–19.9	22.7	15.9	10.3	7.7	4.4	2.5
20–29.9	20.7	16.7	13.6	11.6	7.7	5.5
30–49.9	27.6	28.0	26.9	24.8	20.8	17.2
50–69.9	14.3	17.3	18.9	19.4	19.3	18.2
70–99.9	9.1	12.7	16.3	17.2	20.1	22.0
100–199.9	4.8	8.0	12.0	14.4	19.8	25.2
200+	0.8	1.4	2.0	4.9	7.9	9.4
	100	100	100	100	100	100
Total no. (thousands)	63.8	51.0	43.6	2731	2653	2655

Source: M.A.F.F. June Census results

known to be falling further and further behind that of larger herds (table 1.8). This may be due to a lack of entrepreneurial ability generally on the smaller farms, and in particular a lack of ability to make decisions to control the use of resources.

Average labour input in 1980–81 was 45 hours per cow per year, an 8% reduction over 1976–7. There is a limit to the reduction in labour hours per cow in herds that have modernised their parlours, feeding and slurry removal systems. Scope for further reductions in hours per cow is restricted until further capital investment is made;

Table 1.8 Herd size and average yield (1980–81).

Herd size	1980–81 Average milk yield/cow (litres)
10–20	3997
20–40	4522
40–50	4926
50–60	4965
60–70	5000
70–100	5241
100–150	5442
150–200	5845
200+	5511

Source: M.A.F.F.

such a step has to be carefully considered as very elaborate systems cannot now normally be justified.

It seems that the least hours per cow are worked in herds with over 150 cows, but that parlour type has a major effect on this figure. If we consider 35 hours per cow per year as typical in herds over 100 cows and that a herdsman works 50 hours per week (in six days) for 50 weeks of the year then one man can, on average, cope with 71 cows per year. In addition, this man needs relief for a further 18% of the working year while he takes time off (one day a week plus two weeks' holiday). There are herds that claim to organise their system so that one man plus relief looks after 150 or even 200 cows but they are exceptional and in such cases the herdsmen often do a minimum of outside work (forage conservation, fertilizer application, slurry removal, even feed allocation). These herdsmen are generally paid employees who specialise in the husbandry of the cows; they are often managed or organised by the farmer whose own input in terms of planning, recording and control may be considerable (tables 1.9, 1.10 and 1.11).

Capital invested in dairying

Specialist milk producers now have about £1500–£2000 per hectare

Table 1.9 Labour hours and herd size for three surveys

Herd size	Average labour hours per cow		
	1972–3	1976–7	1980–81
10–20	112	105	126
20–30	87	76	83
30–40	70	63	61
40–50	58	53	53
50–60	51	50	48
60–70	46	44	40
70–100	41	38	37
100–150	} 40	} 34	33
150–200			30
200+	34	35	32
All	60	49	45

Source: M.A.F.F. raised sample results

Table 1.10 Changes in inputs of labour with herd size (1980/81)

Herd size	% of total labour hours		
	Paid	Farmer and spouse	Other family (unpaid)
10–30	1	73	26
30–40	8	83	9
40–50	23	60	17
50–60	13	49	38
60–70	26	53	21
70–100	46	40	14
100–150	68	22	10
150–200	88	7	5
200+	91	5	4
All	35	48	17

Source: M.A.F.F. raised sample results

Table 1.11 A cowman's day (total 10 hrs/600 mins)

	Minutes
Milking	282
Cleaning equipment	64
Feeding	86
Manure removal and handling	74
Littering cubicles	40
Management	54

invested in tenant's capital. This comprises machinery, livestock and deadstock as well as a figure for the valuation for tillages carried out. In 1982 farmers invested an average £100/hectare in new machinery. The return on their investment capital (based on the management and investment income they made) was 14% on average. The top 25% of producers obtained a return of 21%. These figures conventionally assume that the farmer is a debt-free tenant. In actual fact, dairy farmers generally have considerable interest to pay on their loans. Where owner occupiers are involved, the return on their total capital (i.e. including land and buildings) is likely to be only about 5% (table 1.12).

The M.M.B. figures for 1983 show a lower return than this on tenant's capital of 8.8% (table 1.13). In the later set of figures, the average farmer finishes the year with an overdraft £4326 larger than at the beginning despite making a 'profit' of £12 239 (table 1.14). The farmers obtained much more money from long term loans (£8823) and private sources (£1019) in order to invest in extra capital items when their own farm-generated net cash flow (£9487) could not cover the cost. The average overdraft position at the year end was £26 177 out of total liabilities of £64 174 (table 1.15).

The M.M.B. sample described covers owner-occupiers and tenants together, so the composite figure resulting is of limited use. Nevertheless, the trend shows how valuations have just kept pace with the annual increase in liabilities (£7500). Farmers must beware of increasing borrowing like this, particularly when the inflation rate (5%) is at last below nominal (bank) interest rates (12–17%). In the past inflation has helped to pay off loans; in the

Table 1.12 Financial performance: specialist milk producers

	1981	1982	Top 25% 1982
Year	1981	1982	1982
Number of farms	26	26	7
Average size of farms (ha)	91.0	93.0	102.0
	Average £ per hectare	Average £ per hectare	Average £ per hectare
Total farm output	1351.3	1491.9	1814.1
Variable costs			
Feedstuffs	464.5	514.6	613.0
Seed	7.3	8.9	8.5
Fertilizers	76.5	88.5	102.1
Sprays and other crop costs	11.7	11.6	8.7
All other variable costs	113.2	115.1	145.5
Total variable costs	673.1	738.7	877.8
Total farm gross margin	678.2	753.2	936.3
Fixed costs			
Rent and rates (incl. imputed)	81.8	89.1	82.8
Regular paid labour	116.7	129.3	137.4
Unpaid labour	79.1	81.1	72.9
Power and machinery	165.1	176.6	196.0
Other costs	57.7	60.5	91.1
Total fixed costs	500.4	536.6	580.2
Management and investment income	177.8	216.6	356.1
Investment in machinery, live and deadstock and tillages	1403.4	1529.0	1727.4
Return on above capital	13%	14%	21%
New machinery invest- ment	80.0	101.0	122.0

Composition of total farm output %

Crops	4	3	2
Milk	73	74	79
Cattle	18	19	19
Sheep	1	1	0
Pigs	2	2	0
Poultry and other	2	1	0
Total	100	100	100

Source: University of Reading Farm Business Data (1982–3)

Table 1.13 Management and investment income 1982–3

	Year ending spring	
	1982 (£)	1983 (£)
Profit	9127	12239
Plus interest	6641	6643
Depreciation on landlord's fixtures	2357	2426
Less unpaid wages	6760	7402
Rental value of land owned	3320	3480
Management and investment income	8045	10426
Tenant's capital	104403	118372
Return on tenant's capital	7.7%	8.8%

Source: M.M.B. figures 1982–3

future the farmer must look to financing his net capital expenditure by cash generated from the business. Increases in valuation may not be expected to be so good in an industry forced to cut back its production by the introduction of quotas and other devices.

The efficient farmer needs to be able to re-equip his farm every ten years or so. The list in table 1.16 shows where his money has to go over an extended period. The figures include a 20% discount which is generally available at present for new purchases and exclude the smaller items of equipment as well as buildings. Spread over ten years, this amounts to £12745 per year and this figure is,

Table 1.14 Cash flow on M.M.B. survey farms 1982–3

	Year ending spring	
	1982	1983
	(£)	(£)
Profit	9127	12239
Adjust for		
depreciation	+7731	+8344
valuation increase	−4051	−4311
increase in creditors less increase		
in debtors	+49	+165
uptake of long term loans	+7002	+8823
transfer from reserves plus private		
income	+2199	+1019
Cash available to meet commitments	22057	26279
Commitments		
loan repayments	4173	6545
private expenditure	6916	8350
tax	1579	1897
Cash available for reinvestment	9389	9487
Capital expenditure (net)	10604	13813
Cash deficit (increase in overdraft)	1215	4326

Source: M.M.B. figures 1982–3

interestingly enough, very close to the figure of £13813 that farmers of the same size farm (83 ha) actually spent in reinvestment in 1982–3, though they had to borrow to do so.

Every means available to increase cash flow from the farm by improving and intensifying the efficient enterprises and by reorganising the inefficient (like heifer rearing) should be explored to avoid a worsening net worth and a decline into overborrowing and bankruptcy. At all times the monitoring of after-tax, after-interest net cash flow will be crucial.

The efficient use of energy

In a study of thirty-four 'efficient' farms, milk yield was well above average while concentrate use was well below.

The rationing system based on metabolisable energy (M.E.)

Table 1.15 Balance sheet of M.M.B. costed farms 1982–3

	Year ending spring	
	1982 (£)	1983 (£)
Assets		
Freehold of farm	154000	174000
Landlord's fixtures	12672	13514
Tenant's fixtures and machinery	18598	21297
Livestock	67902	76968
Debtors	9267	10703
Crops, stores and tillages	8636	9434
Total assets	271075	305916
Liabilities		
Current		
creditors	10505	12106
bank overdraft	21847	26177
Deferred		
mortgages	6551	6212
hire purchase loans	1751	1776
other loans (private and bank account)	16097	17903
Total liabilities	56751	64174
Net worth	214324	241742
Equity	79%	79%

Notes
The composite balance sheet above includes data from both tenanted and owner-occupied farms. The freehold figures are based on 40 hectares being owner-occupied. The value applied per hectare was the average sale price of land in the spring of 1982 and 1983 respectively.

While the increase in dairy herd valuation is now shown on the trading account it is, as a capital asset, included in the balance sheet. The valuation increase for the dairy herd was £38 per cow and for other livestock £41 per head.

Similarly, the value of home produced forage (hay, silage, etc.) is held constant on the trading account but valued in real terms on the balance sheet.

Equity is net worth as a percentage of total assets.

Source: M.M.B. figures 1982–3

provides an effective answer to the problem of measuring feeding efficiency. The megajoule (MJ) is the feed energy unit and 1000 MJ is known as a gigajoule (GJ). A feed is only used when eaten –

Table 1.16 Capital required for a typical dairy farm

	Item	Value (£)
1	150 new cubicles (excl. installation)	2850
2	16/16 herringbone with A.C.R. (installed)	24500
3	750 gallon bulk milk tank	9995
4	Bulk concentrate bin (14 tonne) and augers	3500
5	50 HP tractor plus front end loader	11118
6	Cattle trailer	1386
7	Yard slurry scraper	215
8	300000 gallon slurry store and pumps installed	21000
9	750 gallon vacuum tanker	2860
10	Fertilizer distributor	1895
11	Six out of parlour conc. feeders and ancillary equipment	9000
12	Mower conditioner 7′ 10″	5585
13	Forage harvester	4895
14	Two 4 foot forage wagons	3920
15	Rotaspreader 5.5 cu yd capacity	1880
16	86 HP tractor	13047
17	Baler	4500
18	Bale trailer 3 ft capacity	995
19	Two tonne roller	700
20	Set of chain harrows	654
21	Hay/silage/fodder/rake	1075
22	Buckrake	635
23	Cattle crush	450
24	20 individual calf pens	800
	Total	£127455

Note
Excludes: extending silage clamp, pasture topper, bale sledge, bale loader, 3rd/4th tractors, fencing materials, installation of cubicles and calf pens, and other minor items.

Farmers Weekly. 16 December 1983

so concentrates are utilised at near to 100% efficiency, while the utilisation of grass depends on conservation or grazing management. When part of a ration known to have come from concentrates or purchased fodder is subtracted from total M.E.

used by the herd, the remainder must have come from forage. The utilised M.E. from forage (U.M.E.) provides a valuable comparison between the levels of output from forage on different farms. A target figure is 100000 MJ (100 GJ) per hectare whereas the average achieved on permanent pasture is only 40 GJ per hectare (table 1.17).

Table 1.17 Milk production from grass: performance of top grassland milk producers 1980–81

No. cows in herd	125
Milk yield litres/cow	5946
Concentrates	
kg/cow	1446
kg/litre	0.24
Nitrogen used kg/ha	338
Stocking rate cows/ha	2.36
*UME gigajoules/ha	104

*Utilised metabolisable energy
Source: Walsh, Rex Patterson Memorial Study, 1982

Utilisation efficiency can be calculated by estimating (from a knowledge of soil type, summer rainfall and nitrogen use) the amount of grass that is eaten (U.M.E.) of that which is grown. In theory, under very good growing conditions using 450 kg nitrogen per hectare, 133 GJ can be grown. A high rate of utilisation is 85% (low 40%) so at the high rate the U.M.E. would be 113.

Although yields were higher at higher levels of concentrate use, margins per cow and per hectare do not show a clear relationship to concentrate input. This may be due to the substitution of concentrates for grass which, without increase in stocking rate, leads to uneconomic performance (table 1.18).

Sensible targets set for modern milk production are 6000 litres of milk per cow using less than one tonne of concentrates (table 1.19).

Prices for outputs in dairying

Recently, major changes have been made in the way farmers are

Table 1.18 Concentrate use, stocking rate and margins in 34 high margin herds

Concentrates tonnes/cow	Stocking rate cows/hectare	Margin over feed costs	
		£/cow	£/ha
Less than 0.9	2.22	610	1182
0.9–1.2	2.27	593	1170
1.2–1.4	2.49	600	1332
1.4–1.6	2.24	597	1213
1.6–1.8	2.34	592	1241
More than 1.8	2.52	636	1458

Source: Walsh, Rex Patterson Memorial Study, 1982

Table 1.19 Targets for profitable milk production

Milk yield litres/cow	6000
Concentrates kg/cow	925
Nitrogen kg/ha	370
Stocking rate cows/ha	2.05
U.M.E. GJ/ha	93
Gross margin	
£/cow	586
£/ha	1201

Source: Milk from Grass, I.C.I.

paid for milk. The producer is now paid at the end of the month in question and only for the weight of fat, protein and lactose that he produces. This has a major effect on returns. Attention to feeding, breeding and health will be well worth while as all these factors affect milk quality to some extent. The other main factors affecting quality are stage of lactation and season of production. In any case, E.E.C. quota restrictions are likely to limit the scope for improving quality economically.

Milk prices are also affected by month of production, where extra payments are made for autumn and winter milk (table 1.20).

Table 1.20 Milk prices per litre affected by month of production

Month	Reduction or addition average price (p/l)	Month	Reduction or addition average price (p/l)
April	− 0.5	October	+ 0.8
May	− 2.5	November	+ 0.8
June	− 2.5	December	+ 0.8
July	Nil	January	+ 0.8
August	+ 1.2	February	+ 0.8
September	+ 1.2	March	+ 0.5

Milk quality measured by total bacterial counts (T.B.C.) also affects price, as does the presence of antibiotic residues detected in the milk.

A knowledge of the factors affecting milk production allows the farmer to decide the calving season and pattern so that he can choose the breeding season for the herd that will lead to highest outputs and margins on the farm. The factors of importance here are soil type, summer rainfall and forage conservation techniques as well as the relative prices of milk, fertilizer and concentrates. (More discussion on this point follows in chapter 2.) Quota restriction may again limit the scope for alterations in patterns of milk production.

Generally, heifer prices have remained static in real terms. This reflects the reduced demand for this stock which shows the farmer's lack of confidence in the future of milk production, as well as the plentiful supply of heifers. Price fluctuations can be quite wild from month to month at particular markets. Subjective assessments of quality in dairy replacements are notoriously unpredictable but not all farmers can avoid using markets by buying from other farmers direct or by having their heifers reared under contract (table 1.21).

On average, heifers in the U.K. are 30.5 months old at calving. They can be more cheaply reared to calve at 24 months. This modern practice releases considerable acreage which can be used for higher margin enterprises; it is unlikely to affect significantly the yield of the herd.

Cull prices are important to fertility management as they affect

Table 1.21 Seasonal price index 1974–83 down calving heifers

	Down calving heifers
January	98
February	99
March	98
April	99
May	98
June	96
July	98
August	99
September	101
October	102
November	104
December	107
Average	100

Source: Bridget Drew, A.D.A.S.

herd depreciation cost. The prices received reflect the weight and condition of the cow but are generally very variable according to demand. Quality improves prices in some cull cow marketing groups, ensuring that better returns are achieved. As cull cow prices average only about half of heifer replacement prices, it is clearly important to reduce culling to an economic level. This should normally be 18–20% of the herd.

Modern fertility management

Interest in fertility management has been growing rapidly. Over the last ten years much attention has been focussed on such factors as heat detection, conception rate, calving patterns and intervals and culling rates. There has been an increasing awareness that herdsmen are hard worked with routine jobs like milking, feeding and slurry scraping and often neglect to carry out the many husbandry tasks that cannot be automated. As automation and mechanisation have taken over some tasks, economic pressures have meant that more cows are purchased and so the time released has been taken up mainly by milking the extra stock. The quest for automatic heat detection has eluded research workers, and good husbandry remains essential for this and other tasks.

About 4.5 million inseminations are carried out in the U.K. each year. The non-return rate quoted for inseminations on the basis of 'non-return' by 30–60 days after service is 78–83%, but the proportion of animals that actually calve to a single insemination is closer to 55%.

This low pregnancy rate, and heat detection rates which also average only around 55%, lead to considerable wastage. The pharmaceutical industry has invested heavily to develop a range of products to help synchronise or predict oestrus and thus to reduce losses due to inefficient heat detection. Unfortunately, these have not been as widely used as had been hoped.

A test for pregnancy by measuring progesterone levels in milk has also been developed. It is being refined at present to make it a test based in veterinary practice premises and precluding the need for expensive equipment. Although manual pregnancy diagnosis per rectum by veterinary surgeons is not widely practised there remains considerable scope for this, perhaps in conjunction with milk tests.

The state of dairy farmers' confidence and profits means that veterinary practices are having difficulty in maintaining their cattle work on farms, and this is giving cause for much concern within the profession. However, properly managed modern veterinary practice has nevertheless been able to cope with reductions in Government work and so-called 'fire brigade' work and to expand through the development of planned animal health and production services. This will be considered in more detail in chapter 6. The following chapters show why fertility management is so important to dairy herd profitability, and how to achieve high levels of performance and ensure minimum wastage.

2 Fertility and profitable milk production

Key measures of profitable dairying

Any financial measure of herd performance should be calculated on an annual basis. The best measures are not only profit per herd, profit per hectare or profit per cow place, but also net cash flow and return on capital. The basic economic performance of the herd can be broadly measured in several ways. These are quick and easy to calculate, but it is important to realise that many other terms should be considered as well in a proper appraisal. The following are some of the parameters which are commonly used.

Margin over concentrates (M.O.C.) is the value of milk sold per cow per annum less the cost of concentrates fed during the year, usually expressed on a per cow basis. It is a measure of performance per cow but does not take stocking rate into account unless M.O.C./hectare is used. All concentrates used, including those home grown, should be included at market values (table 2.1).

Margin over purchased feed (M.O.P.F.) is similar to M.O.C. except that the costs of purchased bulk foods such as brewers grains, sugar beet pulp, molasses, straw or hay are also deducted from the value of the milk sold. Market values should be used for all feed not home grown (table 2.1).

Margin over feed and forage (M.O.F.F.) is defined as the value of milk sold less all feed costs and forage costs. These latter include the costs of seeds, fertilizers, sprays and silage additives used on all the forage area utilised by the dairy cows.

In the 1983 M.M.B., Costed Farms report, M.O.C. per cow averaged £555 per annum and £1110 per hectare. The top 25% of herds had an average M.O.C. of £620 per cow and £1556 per hectare, so the range in performance as measured by M.O.C. is wide and reflects genuine differences in profitability. Measurements on a per hectare basis reflect both cow performance and the

Table 2.1 Output, variable and fixed costs of a typical dairy cow

	£/cow/year	
Output		
Milk sales 5480 litres at 15p/litre	822	
Value of calf (allowing for losses)	72	
Gross returns	894	
Less herd replacement (25% of herd/year)		
heifers at £600, culls sold at £300		
£600 − 300 = £300 ÷ 4 = 75	−75	
Gross output	819	819
Variable costs		
Purchased concentrates 1.6 t at £160/tonne	268	
Forage variable costs		
fertilizer, sprays, seeds	72	
A.I. and bull costs	9	
Veterinary and medicine	18	
Stores and fuel	15	
Dairy cow insurance	3	
Recording fees	5	
Total variable costs	406	406
Gross margin (gross output less variable costs)		413
Fixed costs		
Labour costs	81	
Equipment and machinery		
running costs	92	
depreciation	37	
Interest charges	53	
Property charges		
running costs	46	
depreciation	53	
Sundries	39	
Total fixed costs	401	401
Profit		£12/cow

continued over

At a stocking rate of 2 cows/h		£/cow	£/hectare
Margin over concentrates/cow/forage ha			
	£822 − £268 =	554 =	1108
Margin over purchased feed/cow/forage hectare			
	£822 − £284 =	538 =	1076
Margin over feed and forage/cow/forage hectare			
	£822 − £356 =	466 =	932
Gross margin	£819 − £406 =	413 =	826
Farm profit	£413 − £401 =	12 =	24

stocking rate on the grassland area utilised by the herd (table 2.1).

Gross margin of an enterprise is gross output less variable costs including the allocated variable costs of grass. Gross output is milk and calf sales, livestock valuation differences (numbers and value) over the year, plus cow sales less herd replacement purchases. From this figure are subtracted the variable costs which include the costs of concentrates, purchased bulk feed, forage, veterinary and animal health and a few miscellaneous items like bedding, chemicals and casual labour. The difference between gross output and variable costs is the gross margin which is expressed per cow per year and per hectare per year. Gross margin is a valuable measurement of the performance of individual farm enterprises such as dairying, though is more time consuming to calculate than M.O.C. or M.O.P.F. or M.O.F.F. and is worked out on an annual basis (table 2.1).

Farm profit or loss is total farm gross margin less the sum of the fixed costs incurred (table 2.1). It represents the surplus or deficit before including any charges such as unpaid family labour costs and notional rent. To make M.A.F.F. farm survey figures comparable between farms the profit or loss figures are adjusted to make the results those of a debt-free tenant farmer.

Net farm income (N.F.I.) is farm profit or loss after adding back interest and ownership charges, minus unpaid family labour costs and notional rent. It represents the reward to the farmer and spouse for their own manual labour management and interest on tenant-type capital invested on the farm, whether borrowed or not.

Management and investment income (M. & I.I.) is N.F.I. minus farmer and spouse labour cost plus paid management costs. It represents the reward to management both paid and unpaid and the return on tenant-type capital invested in the farm, whether borrowed or not.

The importance of lifetime performance

As can be seen from the beginning of this chapter, the parameters by which success can be measured in dairying are diverse. Nevertheless, profit per cow, per hectare and per farm is the prime objective, measured by whatever means. This objective may be confused in the minds of some farmers who may see, for example, high milk yield per cow or maximum stocking rate as most important. In the final analysis, however, profitability is the sole yardstick by which success or failure may be measured even for those farmers who see dairy farming as a way of life rather than as a means of livelihood.

Perhaps this view can be summarised by defining the ideal dairy cow as one which produces as much milk per annum over as many years as possible and at economic cost. This embodies the major relationships between input and output and the important fact that lifetime production in physical and financial terms is the key measure of success. The published figures vary but the average life of the dairy cow in this country is around four lactations, giving an annual turnover of 25%.

In the minds of most dairy farmers, lifetime production commences when the heifer calves down for the first time. However, the majority of farmers rear at least some of their heifer replacements and this stage in the life of the cow must also be taken into account when measuring performance. The interval from birth to point of first calving is a non-productive stage and must be considered as an overhead cost to the milking herd. The age of the heifer at first calving is therefore important, and a heifer calving at 24 months not only requires less forage acres from birth to calving than the heifer calving at 36 months but is bringing in money 12 months earlier. In addition, the M.M.B. have shown that the lifetime production of the heifer calving at 2 years is 6% higher than her counterpart calving at 3 years, despite a slightly lower first lactation yield in the younger animal; and the milk produced 'per day of life' is 20% higher! Too often the heifer rearing stage of the dairy cow's life is neglected with the result that the heifers calve in at $2\frac{1}{2}$–3 years of age. Table 2.2 summarises these points.

Table 2.2 Effect of age at first calving of heifers on lifetime performance

	Age of heifer at first calving (months)				
	23–25	26–28	29–31	32–34	35–37
Herd life (lactations)	4.0	4.0	3.8	3.8	3.8
Lifetime yield (kg)	18747	18730	17964	17991	17657
Yield/day in herd (kg)	13.1	13.2	13.1	13.1	13.1
Yield per day of life (kg)	8.8	8.4	7.9	7.5	7.3

Source: M.M.B.

As already mentioned, the lower the age of the heifers at first calving, the less the forage area required to maintain them. The surplus hectares can be released to carry more dairy cows, or for a cash crop, or even to carry extra heifers for sale. Furthermore, the longer the productive life of each cow in the herd, the lower the replacement rate and the smaller the size of the heifer rearing enterprise. This also has an effect, not only on lifetime perform- ance but also on the profitability of the total dairy unit. There are circumstances when the herd replacement rate is deliberately high; for example, when a fast rate of genetic improvement in the herd is a primary goal. However, a high replacement rate for unplanned reasons reduces both lifetime production and profitability. A low replacement rate with heifers calving in at two years of age is most profitable.

The raising of heifers to calve down at 22–24 months of age allows little room for error in rearing or grassland management. Heifers should reach a minimum of 325 kg when they are bred at 13–15 months of age. This demands a daily liveweight gain of 0.7 kg per day throughout this period. Such a rate of gain is fairly easy to achieve in the summer months at grass, but it is essential that it is achieved during the time of breeding to permit maximum conception rates and a concentrated calving pattern. This may demand supplementary feeding with concentrates during this critical period. None of these targets is outside the scope of the good stock rearer, but they do require careful planning and control.

Whilst two year calving of heifers may be the economic ideal, the overall policy for the herd usually demands that heifers calve down for the first time at a specific time of the year. In herds which have a planned calving season, heifers should ideally calve down shortly before or at the commencement of the calving season for the main herd. This ensures:

(a) that the heifer has a reasonable period to recuperate before re-breeding, and
(b) that the animal commences her productive life in the herd at the most profitable time with a reasonable chance of continuing to calve down at that time each year.

Heifers which calve down later than desired in the season are rarely brought forward at subsequent calvings but are more likely to slip back in the calving pattern and end up as barren. Because they calve later they are also likely to be less profitable animals. In the context of two year calving, the ideal heifer is one which calves down both at the desired time and at 22–24 months of age, which means that the heifer herself must have been born at the same time of year two years previously. This can be difficult to arrange.

The annual production cycle

Once the heifer has calved at the optimum time then the annual production cycle commences. The objectives for that animal are firstly to produce the maximum amount of milk from the quantity and quality of feed available to her, and secondly to get in calf again so that she calves at the same time next year. Naturally, the farmer will concentrate on the first objective for daily milk yield is easily measured and changes in management and diet are rapidly and visually reflected in the milk jar at each milking. The second objective is often relegated to a position of minor importance, largely because it relates to the next lactation while the first objective relates to the present.

Turning to milk production first of all, fig. 2.1 shows the annual production cycle of the cow in diagrammatic form. Daily milk production peaks at 35–50 days after calving and then drops at the rate of 1.5–2.5% per week during the remainder of the animal's lactation. The shape of the lactation curve for each animal will

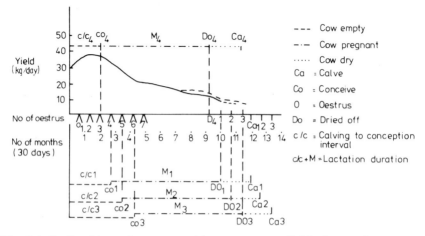

Fig. 2.1 Stylized lactation curve with oestruses available for service showing effect on calving to conception interval, calving interval and lactation length c/c 1,2,3 illustrate the effect of later conception on lactation length (DO 1,2,3)

vary according to parity and month of calving, but the level of production will depend on the quantity and quality of the forage available (whether grazed or conserved) and the amount of supplementary concentrate feeding. Total yield will also increase up to the fourth lactation.

Each annual production cycle begins with calving. It can therefore be argued that the most critical date in the cycle is the date on which she conceives, for this controls her production pattern for the next lactation. The generally accepted optimum interval between calvings (the 'calving interval') is 365 days, firstly because annual milk yield (as opposed to lactation yield) is then at a maximum and secondly because if a heifer calves at the optimum time she should continue to calve in the same period each year. There may be circumstances where heifers are allowed a longer than normal rest period before re-breeding, or where management policy dictates otherwise, but normally a twelve month calving interval should be the target. The point to be made here is that the next calving date for each heifer or cow must be planned soon after the previous calving.

With a mean gestation length around 282 days for the British Friesian cow and a target calving index of 365 days, the animal should conceive as close to 83 days after calving as possible. In practice, most farmers will commence breeding their cows long before 83 days post partum. With the widespread use of artificial insemination (A.I.) in the dairy herd, oestrus must be detected before a cow can be bred. Thus the actual date on which an animal is inseminated is very unpredictable in the absence of some form of oestrus control. The majority of herd managers commence breeding around 42–50 days after calving so that the average interval to first service is 65 days post partum. This in turn should result in the majority of cows conceiving at an average interval of 83 days after calving but with a spread of individual intervals around this mean.

For each individual cow, even under optimum levels of heat detection and conception rate, there will be some variation each year around the calving index of 365 days. The problem animals are those which have not conceived by long after 83 days post partum, and a decision has to be made whether to continue to breed such an animal or whether to dispose of her because she will calve subsequently at an unacceptable time of year or because she will be dry and unproductive for too long a period. This is one of the major reasons for culling dairy cows out of the herd and in many cases it is an unnecessary waste of an otherwise healthy and productive animal.

At the end of lactation cows normally dry off, but high producing cows may have to be forcibly dried off 50–60 days prior to calving and the commencement of the next lactation. Such a dry period is necessary to enable the cow:

(a) to replace body nutrients lost during the previous lactation,
(b) to repair and replace the secretory tissue of the udder,
(c) to allow the unborn calf to develop, and
(d) to help store body reserves for the next lactation.

Cows should ideally build up sufficient reserves to calve down in good body condition; neither over fat, which can lead to liver disorders and acetonaemia, nor too thin, which can result in failure to achieve satisfactory peak yield and poor fertility.

In summary, the lifetime production of the dairy cow commences with the heifer calf born at the right time, calving down at two years of age, and subsequently calving regularly every twelve months and for as many years as possible. Provided such an animal has the right genetic background and is fed for optimum economic yield then she will make the maximum contribution to the profitability of the herd.

The importance of calving patterns

Given that the efficiency of forage utilisation is the key to success regardless of the system of milk production followed, then the optimum calving period must be related to grass growing season and conservation methods. Many factors must be taken into account, but a critical decision must be made as to when the herd should calve and over what length of time. This is related to the relative prices for milk produced in summer and winter, the costs of concentrates and the ability to produce milk from grass and conserved forage. For example, a herd calving in the autumn and producing average yields of, say, 5000 litres per cow will probably be averaging 12–15 litres per cow per day in May after the cows have been turned out to grass (fig. 2.2). With well managed grass capable of producing 22 litres of milk or more at this time of year, the grass is not being used to full potential. In contrast, a strictly autumn calving herd averaging 20 litres of milk per cow daily at turnout in spring is using grass more efficiently.

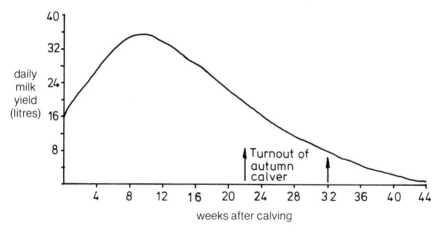

Fig. 2.2 The importance of date of calving as it affects yield at time of turnout

It is important that the farmer has clear objectives for his herd and these should include the period of calving. Far too many herds seem to have no planned calving period on the one hand and, on the other, they combine low stocking rate with high concentrate usage. The potential performance is similar for both autumn and spring calving herds (table 2.3) with less concentrates being used in

Table 2.3 Potential for profitable milk production: implications of season of calving

	Autumn calvers	Spring calvers
Milk yield (l/cow)	6000	6000
Concentrates (kg/cow)	925	465
Fertilizer (kg/ha)	370	370
Stocking rate (cows/ha)	2.05	1.90
U.M.E. (GJ/ha)	92.9	95.8
Grass utilised (%)	77	79
Gross margin/cow (£)	586	623
Gross margin/ha (£)	1201	1183

the latter though with an associated slight reduction in stocking rate.

Spring calving is practised on a minority of farms and here the emphasis is upon achieving a concentrated calving period of about three months and with the majority of the herd calving in January and February. The successful practitioners of this system are all very good managers of grassland. With any 'block' calving system such as this breeding management is absolutely critical, otherwise too many cows will fail to get in calf within the short serving season. Block calving is essential for spring calving herds but it also has certain advantages in autumn and early winter calving herds. Management of the herd is simplified. Routine operations such as calving and breeding occur in a concentrated fashion when farm staff can direct their efforts to major events in the calendar as they occur in natural sequence. Grouping of the cows for feeding is also simplified. As a result, herd performance and profitability can be improved and the situation can be avoided where, for instance, the late spring calving cows in an autumn calving herd often fail to milk for more than 250 days and only have 90 days from calving to be got in calf before culling must occur.

Taking the current U.K. price structure (1985), the effect of that policy on the margin over concentrates for each month of calving can be worked out using factors as shown in table 2.4. Compared with the average price, say 15p per litre, cows calving in, for instance, October will produce a lactation which is worth 4.6% more. The cow calving in that month will also use 1.1% more

Table 2.4 The weighted economics of calving date

Month	1 Milk price	2 Conc. use	3 Yield
January	0.989	1.00	1.009
February	0.962	1.00	0.990
March	0.946	1.00	0.973
April	0.930	1.00	0.961
May	0.949	1.014	0.956
June	0.993	1.098	0.962
July	1.035	1.121	0.986
August	1.046	1.098	1.0135
September	1.046	1.055	1.0285
October	1.050	1.011	1.0465
November	1.039	1.002	1.043
December	1.017	1.000	1.0310

concentrates than average for the same milk production. Compared with the average, cows calving in October will also give 4.65% more milk in a lactation.

Example 1 – cow calving in October

In a herd with a yield average of 5000 litres in 305 days, producing milk of average price (15p/litre).
Herd average concentrate use 0.2 kg/litre at £150/t.

Output
5000 litres × 1.0465 × 15p × 1.050 = £824

Concentrate costs
0.2 kg × 1.011 × 5000 litres × 1.0465 × £150 = £158

Margin over concentrates = £666

Example 2 – cow calving in April in same herd

Output
5000 litres × 0.961 × 15p × 0.930 = £670

Concentrate costs
0.2 kg × 1.0 × 5000 litres × 0.961 × £150 = £144

Margin over concentrates = £526

Difference between two cows £91 (£666 − £526 = £140)

Whereas all lactations are assumed to be 305 days long, in practice the month of calving also affects lactation performance in another way. Cows calving in a particular month will have lactation length decided partly by farmer policy, partly by husbandry and partly by the physiology of milk production. Cows calving earlier in the calving season will not be served until an arbitrary date when the serving season opens. For a cow calving on 6 September in an autumn calving herd this will be about 1 December. Such a cow will therefore be calved at least 90 days before she is served. Assuming typical heat detection efficiency, the cow will be served at about 105 days post partum and will need, say, 1.6 serves per conception adding 20 days before conception at around 125 days. As each day's delay in conception leads to only 0.65 of a day extra on the lactation (beyond 305 days), then the 40 extra days (125 − 85) to conception will lead to 26 (0.65 × 40) extra days on lactation (305 + 26 = 331). This extra 26 days on the tail of the lactation generally produces 10 or so litres per day, thus only some 260 extra litres may be produced. Often cows calving in the autumn and managed this way will milk on a little longer as the end of their lactation comes in the summer when they are usually at grass (table 2.7).

Cows calving after February or March generally have short lactations. They are often served early after calving as the serving season is of limited length (typically ending in June or July) in many herds. The calving to conception interval for these cows is often less than the perfect 85 days, so if their dry period is to be kept at 60 days or so then the lactation length will shorten by a day for each day's reduction in calving to conception below 85. Hence cows first served at 45 days post calving and conceiving at 75 days will have 295 day lactations on average. In addition, many of these animals dry off early because of poor management in late lactation (in October–November) when adequate forage supplies are not made available or when 'low' yielding cows are dried off early by the herdsman who is then preoccupied with coping with freshly calved cows. Often this 30 or so days lost at the end of lactation for late spring/early summer calvers is worth 300–450 litres of milk, which would in fact cost very little more to produce as the animal can produce the amounts involved from good quality forage such as white turnips or a new seeds ley or by opening the silage clamp early rather than relying on wet autumn grass which may not even maintain the animal let alone produce milk.

The importance of calving intervals

We have seen already that the amount and value of milk output, the margin over concentrates and lactation length are affected by month of calving. As we are interested in the production of milk per cow per year, we need to know the optimum calving interval as well as the optimum calving season. Every conception date in one season influences the date of a future calving and hence the date of future conceptions, and so on. Sometimes an annual calving interval will be economically optimal, and sometimes a slightly longer one will have advantages because subsequent calvings may then be at a more advantageous time. These kinds of variables are very difficult to assess without a computer. There may be other factors involved, such as whether it is better to have longer calving intervals and a lower culling rate, or *vice versa*. This clearly affects the age of the herd, the herd depreciation costs and the resources required for rearing heifers. Usually, however, the target is to ensure that individual calving intervals fall between 330 and 370 days.

Using a computer program designed for the purpose, lactation performance targets were set at 4777 litres for heifers, 5460 litres for second lactation, 6142 litres for third lactation and 6370 litres for fourth lactation (giving a 5687 litres average), and the production from forage was assumed to be fairly modest. It was then possible to estimate the cost of each day lost in calving index from a given target.

Assuming an autumn calver extends its calving interval, it will lose 13 litres for each day that it slips beyond 380 days (table 2.5). The cost of a lost day then is £1.36 (table 2.6). The slip in calving pattern is again important because cows calving in certain months have a margin over concentrates that is less than for those calving in others. In the case of cows calving in June it is £95 lower than for those calving in November (tables 2.7 and 2.8). This is due to a shorter lactation as well as poorer milk price, even allowing for different concentrate use.

To our £1.36/day lost profit we can add the cost of the change in

Table 2.5 Effect of increasing calving interval by one day, from 380 days, on annual milk yield (over four lactations starting in month as shown)

Month of first calving	J	F	M	A	M	J	J	A	S	O	N	D
Reduction in annual milk yield (litres)	13.6	12.7	9.5	10.9	10.9	0.9	5.0	5.5	0.9	11.8	20.0	9.2

Table 2.6 Cost of a lost day on a calving interval

	Cost (p)	
Milk not produced 13 litres at 15p/litre	195.0	
Less concentrates not fed less 13 litres fed 0.3 kg/litre = 3.9 kg conc. at £150/t	58.5	
Cost of lost day	136.5	(£1.36)

Table 2.7 Average lactation lengths by month of calving in an autumn calving herd (calving from 1 September to 30 June)

Month of calving	Lactation length (days)	
July/August/September	331	↑
October	320	
November	305	
December	305	Difference
January	305	56 days
February	295	
March	285	
April/May/June	275	↓

calving pattern if the cow slips from November to later months in the season. Over the seven months from November to June the £95 works out at 45p/day, making £1.81 lost/day in total.

In herds where short calving to conception intervals are won by having a high culling rate, it is important to know how much the extra culls are costing (table 2.9). Is it better, for instance, to have a calving to conception interval of 91 days on average and nine cows sold for failing to conceive or 102 days and four culls? (Table 2.10.)

Table 2.8 Estimated margins for each month of calving in typical autumn calving herd

	Yield (1)	Milk value* (£)	Conc. used (kg)	Conc. value (£)	M.O.C.** (£)
August	5526	867	1767	265	602
September	5558	872	1711	256	616
October	5619	884	1660	249	635
November	5749	895	1690	253	642
December	5648	861	1655	248	613
January	5494	815	1588	238	577
February	5559	802	1614	242	560
March	5462	775	1571	235	540
April	5429	757	1563	234	523
May	5443	774	1593	238	536
June	5398	804	1716	257	547
July	5369	833	1741	261	572

* Using factors from table 2.4, column 1
** M.O.C. = margin over concentrates per cow for each month of calving (milk average 15p/litre; concs £150/tonne)

Table 2.9 Cost of culling an extra cow

Cost of a cull	£	£	Difference (£)
Sale price cow culled		350 ⎤	300
Cost of heifer replacement		650 ⎦	
Lower yield of heifer (as opposed to older cow)			
1000 litres at 15p/litre	150		
Less concs 0.3 kg/litre (300 kg at £150/tonne)	45		
	105		+105
Smaller calf 10 kg at £1/kg	10		+ 10
			415

Table 2.10 Comparison of two herds: culling against calving to conception

	Herd A	Herd B	Difference between Herd A and Herd B
Calving to conception	91 days	102 days	11 days longer
Culling for failing to conceive	9 culls	4 culls	5 culls fewer

Herd B Longer calving to conception costs	Lower culling saves
11 days at £1.81 for 96 cows = £1911	5 culls at £415 = £2075

Herd B has a net saving of £164; longer calving to conception interval but lower culling

Table 2.11 Cows culled for each reason, expressed as a percentage of cows calved

Low yield	3.8
Infertility	7.6
Mastitis	3.2
Behaviour	0.5
Old age	0.7
Death	1.8
Injury/accident/lameness	1.8
Unclassified	3.2
	22.5

Source: Daisy – The Dairy Information System, University of Reading

It may be preferable to have a lower culling rate paid for with a longer calving to conception interval. It is all too common to see extra culling carried out wastefully (table 2.11). Total culling rates should not exceed 18–20% per year for optimal performance as this usually means that where heifers are reared at home the minimum acreage is allocated to replacements and the maximum to dairy cows. Every extra heifer takes 0.4 ha to rear to two year calving and this means that a lower gross margin (£800 − £300 = £500/ha) of £200/heifer has to be borne by the farm.

Interaction of heat detection, pregnancy rate and culling rate

There is a relationship between interval to first service, heat detection and pregnancy rate which can be described in terms of calving to conception intervals and culling rate. The culling rate is worked out on the basis of cows failing to get back in calf by 190 days after calving. Table 2.12 shows the position in herds that calve around the year. Where pregnancy rate is reduced from 60% to 50% and where heat detection stays at say 60% then about four more culls are sold for failing to conceive and about another five days are lost on the calving to conception interval.

Table 2.12 Effect of heat detection, pregnancy rate and interval to first service on calving to conception and culling rate

Average interval to first service (days)	Heat Detection Rate					
	50		60		70	
	Conception Rate					
	50	60	50	60	50	60
85	107/12	102/7	107/8	102/4	106/4	102/2
75	99/11	93/6	98/7	93/3	97/3	90/2
65	91/9	85/5	80/6	84/3	88/2	81/1

Note
First figure is calving to conception in days (e.g. 107) and second figure is percentage of cows failing to conceive by 190 days post partum (e.g. 12%).

Bringing these facts and figures together in terms of one herd of 100 cows, the effect of poor fertility management can clearly be disastrous (table 2.13). The overall effect on the milking herd alone can amount to £101/cow/year when heat detection (50%) and pregnancy rate (50%) are poor as opposed to good (80% and 60%

Table 2.13 Effect of poor fertility management on dairy herd profitability

	Poor herd	Good herd	Difference	Cost/cow (£)
Heat detection (%)	50	80	30	
Conception rate (%)	50	60	10	
Interval to first service (days)	85	65	20	
Effect in 100 cow herd				
Calving to conception interval (days)	107	81	26 × 1.81	45.06
Culling rate failing to conceive (%)	12	1	11 × 415	45.65
Cost of semen (£) 10% difference in pregnancy rate ($\frac{1}{2}$ nominated, $\frac{1}{2}$ bull of the day)			3.85/cow	3.85
Extra vet. costs time mins/cow/yr (5 mins at £30/hr)	12	7	2.50/cow	2.50
Extra treatment/cow (£)	12.50	9.0	3.50/cow	3.50
				102.56

respectively). The cost of poor fertility management in dairy heifers is also very significant and should be added to this total, but at this stage in a 100 cow herd we are already talking about £10 256 lost profit per herd per year.

The most realistic way of measuring the economic performance of the dairy enterprise as a whole is to calculate the margin per hectare from the combined area of the farm devoted to both dairy cows and dairy youngstock. This avoids the situation where two-thirds to three-quarters of the grassland area is apparently producing a high margin per hectare from the dairy cows while the remaining area carrying the dairy heifer unit is conveniently forgotten.

If the youngstock unit is intensified by increasing the fertilizer usage in order to improve the stocking rate then fewer hectares are required for dairy heifer rearing. However, a more dramatic effect is achieved by reducing the age at calving from three to two years, as has already been discussed. Assuming the stocking rate remains constant at 1.9 livestock units per hectare, then the area required per heifer from birth to calving is reduced from 1.0 for three year calving to 0.55 hectares for two-year-old heifers. On the 70 hectares previously carrying 100 cows and producing 25 down calving heifers per annum the effect is to release more land for dairy cows or other enterprises. Assuming more cows can be carried, then at the same stocking and replacement rates as before the changes in gross margin can be calculated (table 2.14).

Table 2.14 Effect of two versus three year old calving on farm gross margin

	3 year calving	2 year calving
Physical		
Area available	70 ha	70 ha
Area available for dairy cows at		
2.25/hectare	45 ha	53 ha
No. of cows	100	120
Area available for youngstock	25 ha	17 ha
No. heifers calving per annum	25	30
Financial	£	£
Gross margin from dairy cows		
at £1110/hectare	49950	58830
Gross margin from youngstock	8750	10810
Total gross margin	58700	69640
Overall gross margin/hectare	839	995

It is the best alternative use that should be considered in decision-making, and on many herds the choice is either to keep fewer heifers and more cows using fewer concentrates for the cows or to grow another high gross margin crop such as wheat on the released land.

Another point worth making is that the herd replacement rate controls the size of the dairy heifer unit, and hence influences overall profitability as well. As will be discussed in chapter 3, breeding management in both the dairy cows and the maiden heifers has an important bearing on the subject.

The date of arrival of the heifer in the milking herd is crucial to her lactation length and lactation yield. Good fertility management of the maiden heifer (good heat detection and pregnancy rates) can lead to 30 days reduction in calving age even for the heifers born before the end of October. In this case every extra day can lead to about 17 litres more milk per day for the extra 30 days, or £1.76 per day margin (17 litres at 15 pence per litre less 5.3 kg concentrates at £150/tonne: £2.55 − £0.79 = £1.76 (table 2.15)).

Table 2.15 Benefits of good fertility management in maiden heifers

In a 100 cow herd 25 maiden heifers/year	
Earlier conception	
30 days at £1.96/day × 25	1470
Earlier calving	
30 days at £1.81/day × 25	1357
Earlier delivery of heifer calves for rearing (8 calves more born before Oct. 31)	2916
	£5743

These potential savings in the heifer unit (£5743) in a 100 cow herd can be added to the £10000 that the dairy cows lose with poor fertility management as well. Clearly this £15000 or so represents a remarkable amount of profit to be gained by tackling the task of getting cows seen, served and saved as effectively as possible.

3 Reproductive physiology of the cow

At the moment when a cow is served or inseminated a large number of factors are interacting to influence the outcome. It is necessary to understand some of these in order to understand what might be happening when fertility is not adequate. The intention in this chapter is therefore to describe the events in the reproductive tract of the cow throughout the breeding cycle and to try to identify the points at which reproductive failure is most likely to occur.

Anatomy

Figures 3.1 and 3.2 show the anatomy and anatomical relationships of the bovine uterus and ovaries and fig. 3.3 is a photograph of a bovine ovary taken from a cow slaughtered on the day of her oestrus.

The uterus of the non-pregnant cow or heifer lies curled within the pelvis or, in older cows, just over the pelvic brim. The urinary bladder lies below the cervix and body of the uterus while the rectum lies above it. The close parallel relationship of the rectum and the reproductive tract is of course a critical feature of veterinary work in bovine fertility. Manual palpation by the veterinarian with his arm inside the rectum provides him with a method of examining the cervix, uterus, both ovaries and other pelvic structures through the rectal wall and forms the basis of most of the routine fertility work described in chapter 6.

The cervix is a thick walled tube surrounding a narrow and irregular space or lumen which when tightly closed, as for example during pregnancy, forms a barrier between the uterine environment and the outside. It opens *only* during oestrus or heat to allow the passage of spermatozoa and at the end of pregnancy to allow the passage of the foetus. The uterine horns are separated from the ovary by the very narrow fallopian tubes which open at their far

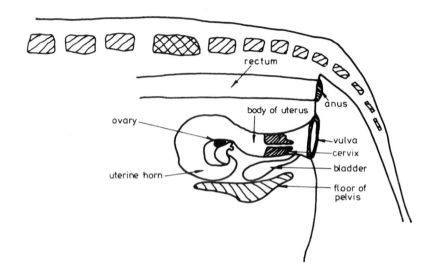

Fig. 3.1 Lateral view of bovine reproductive tract to show the anatomical relationships with other structures

end into the funnel shaped *fimbria* (see fig. 3.4) which are each, in turn, closely associated with an ovary but not connected to it. This 'bursa' or pouch of tissue helps to guide the egg from the ovary to the opening of the fallopian tube.

In summary, the vagina is separated from the body of the uterus by the cervix. The uterine body divides to form two uterine 'horns' which become narrower at their far ends, eventually becoming the very fine fallopian tubes. These end in a funnel-shaped structure close to the ovary, ready to catch an egg when it is released (see fig. 3.4).

The heat cycle

What happens in the reproductive tract

On the day of heat a large fluid-filled sac, the 'follicle', is to be seen on one and occasionally on both ovaries. This follicle contains the maturing egg or ovum (a single cell) which is surrounded by a cluster of smaller cells and bathed in the follicular fluid. The follicle also secretes from its lining (or 'thecal') cells the hormone *oestrogen*

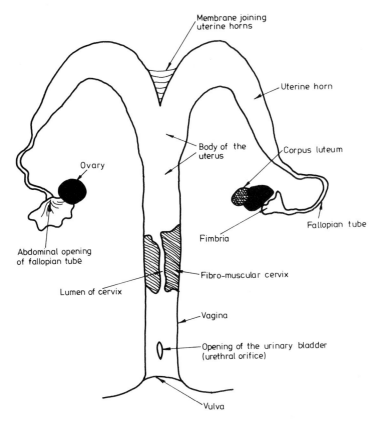

Fig. 3.2 Dorsal view of the reproductive tract of the cow with the uterus 'uncurled'

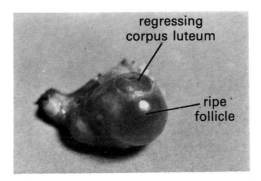

Fig. 3.3 Bovine ovary taken from a cow slaughtered on the day of her oestrus

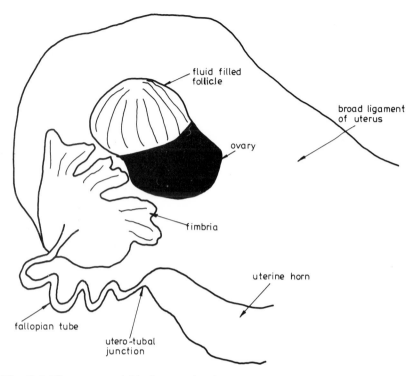

Fig. 3.4 The ovary within its ovarian bursa

(more accurately, oestradiol–17β) supplies of which, for 2–3 days before oestrus and on the day of oestrus, are responsible for a wide range of physiological changes including:

(a) An increase in blood supply to the uterus.

(b) Increase in the 'tone' or firmness and activity of the muscular wall of the uterus.

(c) Alterations in the nature and number of the glandular cells in the uterine lining (the 'endometrium').

(d) Relaxation and opening of the cervix and considerably increased secretion of cervical mucus which is essential for the efficient transport of spermatozoa through the cervix and into the uterus. This mucus is often seen outside the vulva and on the flanks and tail during and around the period of heat.

(e) Perhaps most importantly, this follicular oestrogen circulating in the blood supply to the brain leads to alterations in behaviour and the obvious signs that the animal is in heat, e.g. standing to be mounted by other cows.

All these changes are designed to ensure that the cow is mated and that conditions in the vagina, cervix and uterus are optimal for the survival and transport of spermatozoa and the fertilization and survival of the egg. It is thought that the homosexual behaviour of cows at the time of oestrus evolved to help signal to a distant bull in the wild that the cow was fit for service.

Through the secretion of oestrogen, the follicle is also responsible for its own rupture or 'ovulation' shortly after the end of heat. When the concentration of circulating oestrogen becomes high enough it stimulates the sudden release of large amounts of the hormone 'luteinizing hormone' (LH) from the pituitary gland at the base of the brain. This so-called 'L H surge' results, 24–28 hours later, in the breakdown of the follicular wall, leakage of follicular fluid and collapse of the follicle; and eventually in the escape of the ovum into the abdominal cavity but inside the ovarian bursa, from where it is rapidly drawn into the fallopian tube by the action of *cilia* at the fimbria and in the tube itself. These cilia are very fine hair-like structures which vibrate and move spontaneously.

The hormonal and physiological changes during the oestrous cycle are shown diagrammatically in fig. 3.5. Those changes occurring between heat and ovulation are an extremely consistent and well defined sequence of biological events, the significance of which will be discussed in more detail later.

What can go wrong – and some treatment

Even at this early stage in the description of reproductive events, it is worth noting one or two things which can go wrong.

1. Inadequate oestrogen production by the follicle may lead to a failure of the appearance of behavioural symptoms of heat, or at least to a much shortened heat period.
2. More significantly, it may fail to trigger the pituitary LH surge. Ovulation may then be considerably delayed until some days after heat behaviour has ceased or it may not occur at all, in which case the mature follicle may simply degenerate and shrink up (become atretic) or it may grow to form a large fluid-filled ovarian cyst. Cystic ovarian degeneration, as it is called, is a well known cause of infertility in dairy cows although its incidence tends to be higher in some areas of the U.K. than in others. Generally, in any year only about 2% of cows are thought to have cystic ovaries. In its

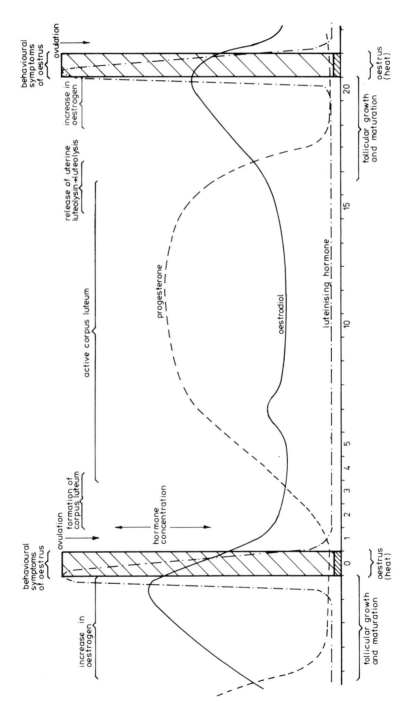

Fig. 3.5 Diagrammatic representation of alterations in circulating hormone concentrations and their relationship with other events of the bovine oestrous cycle

early stages the ovarian cyst tends to be thin-walled and produces large amounts of oestrogen. This causes animals to experience repeated bouts of heat behaviour at short intervals (every 2–3 days) or even to remain in heat all the time. Subsequently the cyst *may* become luteinized (this term is explained more fully later on), when it will produce progesterone. At this stage the cow cannot show heat at all and she may enter a period of prolonged anoestrus. Ovarian cysts are more common in some breeds (e.g. Jerseys) than others and, as already mentioned, in some areas (e.g. south west England). They are not difficult for the veterinary surgeon to diagnose by rectal palpation, and they will generally respond to available treatments such as prostaglandins or compounds with LH activity.

The so-called 'holding injection' used by the veterinary profession and given on the day of insemination is either human chorionic gonadotrophin, which has luteinizing hormone activity, or, more recently, synthetic gonadotrophin releasing hormone (Gn RH) which stimulates the cow's own LH surge. These treatments are designed to prevent the possibility of delayed ovulation but it is difficult to know how often they are effective.

What happens at service or insemination

The significance of delayed ovulation to the farmer is that by the time the ovum is released the spermatozoa in the tract are no longer capable of fertilizing it. Fresh sperm deposited by a bull in very large numbers in the anterior vagina pass through the cervix in strands of cervical mucus and are then carried by the motility of the myometrium (uterine musculature) very rapidly to the *uterotubal* junction (see figs 3.2 and 3.4). From various observations in the bovine and other species it is generally agreed that sperm must be in the female tract for a period of 6–8 hours before they are capable of fertilizing an ovum. The changes which take place in the sperm head (acrosome) during this period are not well understood but they are certainly essential and are known under the general heading of 'capacitation'. Deep frozen semen, thawed and deposited by the inseminator through the cervix and into the body of the uterus, also requires this capacitation time and is transported up the tract towards the ovary in the same manner as fresh semen. Total sperm survival time in the tract is thought to be

24–48 hours in the bovine, so excessive delays in ovulation may obviously result in failure of fertilization.

Ovulation timing

As already pointed out, the trigger for release of the ovum is the pituitary surge of luteinizing hormone which occurs within 2–3 hours of the onset of standing heat. Bearing in mind that:

1. standing heat may last more than 12 hours,
2. LH release occurs shortly after the onset of standing heat,
3. ovulation takes place 24–30 hours after the onset of heat, and
4. sperm need 6–8 hours for capacitation in the tract,

it can be deduced that the best time for service or insemination is towards the end of heat and about 12 hours before ovulation. However, although this assumption is generally borne out by observations in the field, it should be emphasized that such timing is by no means critical. It is also fair to say that if it were critical then the daily system of artificial insemination would not be as successful and well established as it is today. As the sperm lasts longer than the egg it may be wise, if in doubt, to serve earlier in the standing oestrus phase, serving a second time if the cow is still standing to be mounted on the following day.

Fertilization and what can go wrong

The final approach of sperm and egg is probably brought about partly by ciliary action and partly by chemotaxis (chemical attraction). Certainly fertilization finally occurs in the fallopian tube or oviduct and 2–3 days later the developing embryo is taken by cilia down into the uterine horn where it continues its development. Provided that sperm and ova of the correct age and viability are available, then fertilization itself does not seem to be a problem. (Studies involving embryo transfer have shown that the success rate of fertilization is probably in excess of 90%, although it is difficult or impossible to design and carry out studies in the normal bovine to examine this phenomenon.) Neither should there be any problem with the passage of an ovum into and down a fallopian tube. However, the lumen of the tube is extremely narrow and it is likely that quite a high proportion of so-called 'repeat breeders' arise because blockage of the tube prevents the transport of the egg and sperm. Retention of the placenta (the 'cleansing') after calving frequently leads to infection and inflammation of the uterine lining (endometritis) which may spread up the uterine horns and affect the fallopian tubes. Even when such inflammation has regressed, the healing processes may leave scarring and fibrosis of the tubes and blockage of the canal. These changes are not easy to diagnose by rectal palpation unless they are very severe.

It is also worth noting at this point that many cases of mild post partum endometritis regress without treatment when the cows come into heat. It is well known to be very difficult if not impossible to infect a bovine uterus which is under oestrogen domination, while it is comparatively easy to infect in the presence of progesterone or in the absence of any particular hormone domination (as in the early post partum cow).

The remainder of the heat cycle – what is happening in the tract

Formation of the corpus luteum

Ovulation itself is a more traumatic occurrence than perhaps is generally realized. The gradual breakdown of the follicular wall and subsequent escape of ovum and follicular fluid is accompanied by quite extensive haemorrhage (bleeding) in the substance of the ovary and the cavity remaining after ovulation soon fills with blood to become what is known as the 'corpus haemorrhagicum'. At this point it can be more clearly understood how luteinizing hormone got its name, for the LH surge is not only responsible for ovulation of the follicle and a sharp fall in the concentration of circulating oestrogen but it also programmes the granulosa cells lining the follicle to undergo the process of 'luteinization'. These cells become highly specialized and invade the corpus haemorrhagicum which soon also becomes organised into these 'luteal cells', and the whole becomes the corpus luteum (or 'yellow body') which grows and develops beyond the surface of the ovary and secretes large quantities of the hormone progesterone. Photographs of some corpora lutea are shown in fig. 3.6. Quite early in its development the corpus luteum develops the rich and complex blood supply necessary both for the nourishment of this very active gland and for the removal and distribution of the products of that activity. The corpus luteum begins to secrete progesterone in concentrations that can be measured in the blood plasma as early as 3–4 days

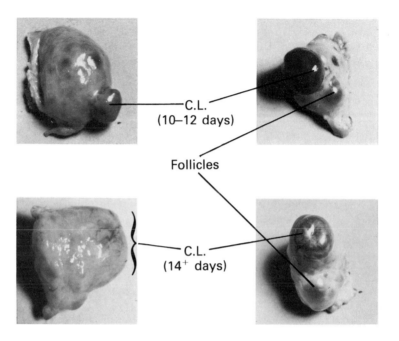

Fig. 3.6 Corpora lutea on ovaries

after the end of heat and continues to do so for about 16–18 days (see fig. 3.5), reaching a peak at around days 10–12 of the oestrous cycle.

Luteolysis

The cessation of progesterone secretion occurs abruptly in the non-pregnant cow and is brought about by the release of the so-called 'uterine luteolysin' and its direct passage from the uterus to the ovary where it initiates regression or degeneration of the corpus luteum and almost immediate cessation of the production of progesterone. This happening is known as 'luteolysis'. Regression is evident in a rapid hardening, shrinking and loss of colour of the corpus luteum as it loses its blood supply (see figs 3.3 and 3.6), and can be detected by repeated rectal palpation. The reduction in size continues as follicular growth and maturation is occurring elsewhere in the ovaries until, on the day of heat, the old corpus luteum is just visible on the ovarian surface (fig. 3.3) and may even be felt by rectal palpation as a small hard area of tissue.

Functions of progesterone

If fertilisation has occurred and a live, developing embryo is present in the uterus, then the corpus luteum does not regress but continues to secrete progesterone and remains active throughout pregnancy. Progesterone, like oestrogen, also has a wide range of physiological effects throughout the reproductive tract including alterations to the number and nature of endometrial glands (endometrial proliferation), closure of the cervix, a reduction of myometrial activity and uterine tone, and in general creating within the uterus the ideal environment for the developing embryo. Certainly normal pregnancy will not proceed in the absence of adequate supplies of progesterone, and failure of luteal function may well be one of the causes of early foetal loss.

It is important to remember also that in the non-pregnant cow follicular maturation, oestrogen secretion and oestrus itself are inhibited by circulating progesterone.

The ovarian cycle as a whole – growth of a new follicle and ripe egg

Luteal regression is followed by the growth and maturation of a follicle which secretes increasing quantities of oestrogen which in turn, after 3–4 days, initiates LH release from the pituitary gland, ovulation, the formation of another corpus luteum and the establishment of another oestrous cycle. The series of ovarian events accompanying and following luteal regression follows a fairly consistent time course, and this phenomenon forms the basis for the success of all currently available methods of oestrous cycle control as well as A.I. itself. (This concept will be discussed more fully in chapter 4.)

Figures 3.5 and 3.7 show diagrammatically the sequence of events occurring in the ovary and elsewhere during the oestrous cycle, and the various hormone changes which are known to accompany these events. The whole cycle takes about 21 days, although the 'normal' cycle length is known to vary between 17 and 25 days. For convenience it can be divided into the luteal phase of 16–18 days length (when the corpus luteum is active) and the much shorter follicular phase which takes place after luteolysis has occurred and includes the maturation or ripening of a follicle, the appearance of behavioural heat and ovulation itself. Occasionally, through failure of proper luteal function, cows may show a

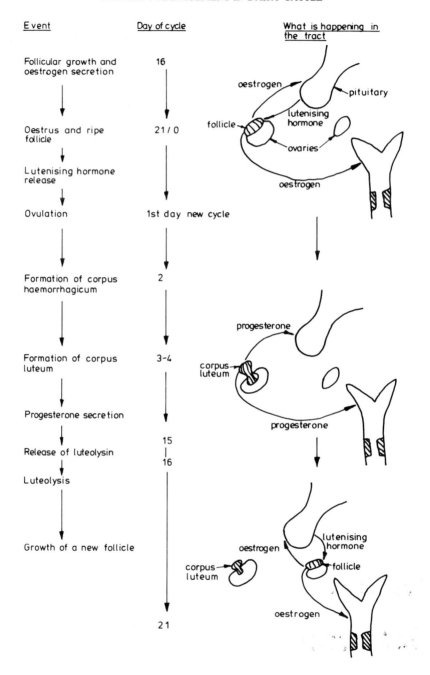

Fig. 3.7 Diagrammatical representation of the oestrous cycle

much shorter interval between heats. This occurs particularly in older cows and in large herds. These 'short cycles' can be followed by ovulations of perfectly normal fertility.

In the absence of successful mating or insemination, and therefore of pregnancy, this ovarian cycle of events should continue to repeat itself regularly every 21 days or so. Failure of this 'cyclicity' is a common problem and is discussed more fully later.

Some problems which can occur

Pregnancy and its failure

Most available experimental data in cattle and other species suggest that fertilization is usually a successful procedure. This means that if insemination or service takes place at the appropriate time then conditions and secretions in the tract will usually ensure survival of the sperm and the ovum and the opportunity will usually be available for the two to meet. So what happens to all the pregnancies which should result? If fertilization is successful on, say, 90% of occasions and pregnancy rate in dairy cows is considered normal if it is between 50% and 55%, then obviously many pregnancies are being lost.

It has been suggested, with considerable supporting evidence, that a proportion of these losses are due to what can be termed 'natural biological wastage'. This results from the fusion of a sperm and egg which are genetically incompatible and are only able to develop for a few early cell divisions before dying. Certainly, evidence from other species shows this to occur and also shows that ageing of either sperm or egg before fertilization in the tract may increase its likelihood.

It should be remembered that venereal infections such as vibrio foetus, although now uncommon in the U.K., are widespread in some parts of the world and can cause large numbers of foetal losses.

The proliferation of the endometrial glands under the influence of progesterone and the increased activity of these glands leads to the secretion of the so-called 'uterine milk'. The function of this fluid is to nourish the embryo during the early stages of its development before implantation or attachment to the uterine wall. The early dividing embryo (blastocyst) emerges from the fallopian tube into the uterus surrounded and protected by a sac (the zona pellucida) which itself offers some nourishment to the blastocyst. The bovine embryo emerges (hatches) from this sac on about the 10th day of development and is then reliant for nourishment upon the uterine fluids. Recent evidence shows that attachment to the endometrium begins to occur at about the 22nd day of life and definite signs of the formation of a placenta are evident by the 30th day. Evidence from many species shows that the hormonal status of the mother is very important if all these changes are to occur properly and if the embryo is to remain viable and continue normal development. Inadequate supplies of ovarian steroids or failure of the developing embryo itself to produce its own stimuli can lead to early embryonic loss.

Recent evidence has shown that most early embryonic loss occurs at about 25–30 days of pregnancy. At this stage the embryo or foetus is still very small and its surrounding membranes are not well developed. Although actual abortion may occur in some instances, it is more common at this early stage of pregnancy for both embryo and membranes to be reabsorbed by the cow. While such reabsorption is taking place the uterus may be incapable of generating the uterine luteolysin, so that the corpus luteum is maintained on the ovary and the cow may not come into heat for some weeks. This obviously results in a considerable loss of time and has been identified as a major cause of infertility at the present time. The problem is currently poorly understood, although it has been shown to occur to a greater or lesser extent in most herds and especially in older cows.

The cotyledons, through which contact is maintained between mother and foetus, must obviously function properly from their formation until the end of pregnancy. Interference with this function is most likely to come from infections (the commonest being brucellosis) at the placental junction, which will result in abortion. Fortunately in the U.K. brucellosis is no longer the scourge that it once was, but it still occurs and in some parts of the world it remains endemic. Other bacterial infections will have a similar effect, and fungal infections of the placenta leading to abortion are becoming increasingly common. It should be clearly understood that the feeding of mouldy or fusty forage to pregnant stock can be dangerous.

Heat cycles after calving (post partum)

The corpus luteum of pregnancy degenerates just before calving, allowing the normal processes of parturition (calving) to occur after which there is a period of ovarian inactivity or anoestrus when heat and ovulation do not occur. Resumption of the normal regular (approximately 21 day) cycle of ovarian events usually follows the first post partum ovulation, which may or may not be accompanied by heat and should occur within about 21 days of calving. If the uterus is undamaged and not infected after calving and if the nutritional status of the cow is adequate then she should be ready for breeding at about 50 days post partum (see chapter 6).

However, it is not unusual for a cow to fail to cycle or to have irregular heat cycles or heats interspersed with periods of ovarian quiescence. Uterine infections and luteinized ovarian cysts ('luteal cysts') can certainly cause the problem, but in the modern dairy cow it is probably more usual for the fault to lie in the feeding. The relationship between weight loss and gain after calving, milk production and dietary intake is complex, and unless it is correct the bovine is very adept at 'switching off' reproductive function. This may be because the quantity and quality of the diet is insufficient to maintain production and body condition, or it may simply be that the cow is not eating enough. Under modern management conditions this is very likely to occur, for instance, in lame cows which cannot walk to and compete at the feed face or reach the grazing areas quickly enough.

If the proportion of cows in a herd which fail to cycle properly is low (less than 10%) then the problem should not have a significant effect on the breeding programme. However, a large number of cows not seen in heat after calving may be indicative of a more deeply rooted management problem which may demand a full investigation by the veterinary surgeon.

Pregnancy diagnosis

The importance of regular and accurate diagnosis of pregnancy in the modern dairy herd is emphasized repeatedly in the course of this book. Without it there cannot be any real control of herd fertility.

Palpation

The most reliable method is still rectal palpation by a veterinary surgeon, some of whom can make an accurate diagnosis as early as 35 days post insemination.

In later pregnancy the calf itself, as well as the gross enlargement of the uterus and the cotyledons, can be palpated and at 5–6 weeks post insemination the slight swelling in the uterine horn and the foetal sac filled with fluid can be felt through the rather thin uterine wall.

Most veterinarians will charge on an hourly basis for this kind of service, so if sufficient help and adequate facilities are available this may well be the least expensive method. The disadvantage of palpation is that veterinarians vary in their ability and willingness to make the diagnosis in early pregnancy.

Progesterone assay

Another method currently available in the U.K. is the assay of progesterone in milk on the 24th day after insemination. Progesterone secreted by the corpus luteum of early pregnancy passes into the milk from blood plasma. As already explained, if fertilization has occurred and a pregnancy is becoming established then regression of the corpus luteum does not occur and progesterone production continues at a high level. This progesterone is measured in milk at day 23, a time when in the non-pregnant cow it would be absent (fig. 3.5). The progesterone concentration is measured very accurately by radio-immunoassay. Absence of progesterone at this time is a 97% certain diagnosis of non-pregnancy. Positive diagnoses, however, are not so accurate since cows undergoing early foetal loss are not diagnosed by this method and abnormalities in oestrous cycle length can lead to inaccurate or misleading results. Nevertheless, positive pregnancy diagnosis is probably 80–85% accurate by this assay and backed up, as all methods should be, by adequate heat detection it should yield good results. The ideal system would be monitoring by milk progesterone assay backed up by confirmation of positive assay results by later rectal palpation; this remains the most cost effective method but it is not easy to convince farmers of the fact.

Oestrogen assay

In later pregnancy (around 4–5 months) the cow begins to produce large quantities of oestrogen which also get into the milk. A milk oestrogen assay is now available in the U.K. for the later confirmation of pregnancy (the oestrone sulphate test).

Heat detection

Most dairy farmers in the U.K. still do not use any method of pregnancy diagnosis other than their routine observation of returns to service after insemination or mating. It is emphasized repeatedly elsewhere in this book that the efficiency of heat detection is extremely variable, and since it is difficult to measure this efficiency over any particular period of time it is also difficult to maintain an assessment of the pregnancy status of the herd. Considerable improvements in heat detection can be effected by the use of aids such as tail paste, vasectomised bulls or Kamars (these aids are described in chapter 5), and if heat detection is being relied upon for pregnancy diagnosis such aids may well be considered essential.

A further problem lies in the very production of oestrogen which forms the basis of the milk oestrogen assay. A proportion of cows produce so much of this hormone in mid and late pregnancy that they show every appearance of being in heat. They attract the attention of other cows and will often stand to be mounted or even to be served by the bull. Where heat detection alone is used to detect pregnancy such cows are often inseminated. Penetration of the cervix by the inseminator's pipette will result in abortion of the foetus, which is obviously a very expensive mistake leading to considerable loss of breeding time.

Other methods

Recently, various pieces of ultrasonic equipment have become widely available to the farming community and are marketed with the promise that the herdsman can at last carry out his own very accurate pregnancy diagnosis. There is little doubt that in skilled hands such claims may prove to be well founded. However, there is equally little doubt that to achieve a high level of skill demands a great deal of practice and constant application – two things which most herdsmen are likely to lack. The technique may be accurate but may not be any easier to perfect than the skill of rectal palpation.

No matter what method of pregnancy testing is used, a small proportion of animals diagnosed as in calf will turn out at full term to be barren. (The proportion of the herd which do this should be well below 5%.) The reason is presumably that abortions which have not been seen have occurred in early pregnancy, and the animal has either not subsequently come into heat or at least has not been seen in heat. There is of course a natural tendency not to monitor heat in cows which are assumed to be pregnant.

The future of pregnancy diagnosis may well lie in the isolation and identification of the so-called 'foetal message' by which the cow knows she is pregnant and which initiated all those physiological changes which are compatible with the continuation of the pregnancy. Recent evidence suggests that such a message can be detected as early as a few hours after successful fertilization and no doubt, in time, there will be a simple method of detecting it.

Similarly, an on-farm colour change method of identifying progesterone in milk will be developed in the near future and this may be cheap enough to use repeatedly on individual cows until pregnancy is confirmed. It could even be automated.

Summary

Possible causes of functional infertility can be summarised thus:

- Failure or delay of ovulation.
- Development of ovarian cysts.
- Failure to exhibit proper behavioural symptoms of heat.

The above are likely to be caused by failure to produce and release adequate supplies of pituitary and ovarian hormones at the right time.

- Infection of the uterus – damage to the endometrium.
- Blockage of the fallopian tube.
- Failure of fertilization.
- Failure of the embryo to develop.
- Early foetal loss/resorption.
- Abortion during later pregnancy.

- Failure of cows to start cycling after calving.
- Failure to maintain regular ovarian cycles.

In this chapter we have attempted to outline the basic reproductive physiology of the cow, and to emphasize those points at which it can go wrong to the farmer's cost. It may be that in doing this we have given the impression that physiological aberrations and functional abnormalities like uterine infections are very common. In most herds this is not the case; by far the most common cause of reproductive failure is failure to observe the cow in heat and/or failure of the heat to occur. Post partum anoestrus remains the main difficulty, and its intimate links with sound nutrition cannot be over-emphasized.

4 Control of the oestrous cycle

Physiological principles underlying cycle control

In order to understand the methods currently available for controlling the oestrous cycle in cattle it is necessary to remember some of the salient points from chapter 3:

(a) While the hormone progesterone (or in fact any substance with a similar activity) is circulating in a cow's blood she will not be able to produce a mature follicle, come into heat and ovulate.
(b) The events following the withdrawal of progesterone from the circulation are a predictable, well defined and well synchronised series of happenings.
(c) These events are:
    (i) The final maturation of a follicle containing the ripe egg (ovum).
    (ii) Increased production of oestrogen from that follicle.
    (iii) The appearance of behavioural symptoms of heat.
    (iv) The release of luteinizing hormone (the ovulating hormone) from the pituitary.
    (v) Ovulation, release of the ovum and the commencement of a new oestrous cycle.
Events (i) to (v) usually take 3–5 days to complete, but events (iv) and (v) are consistently separated by about 28 hours (fig. 3.5).

It follows from these points that there are two logical approaches to the control of the bovine oestrous cycle:

(a) To apply an 'artificial' corpus luteum (or source of progesterone) for a specific period of time.
(b) To control the life span of the cow's own corpus luteum (the origin of the progesterone).

Progesterone/progestational treatments

All the earlier attempts to synchronise heat were based upon approach (a), and the first (and perhaps simplest) of these upon the daily injection of a dose of progesterone in corn oil, known to be sufficient to inhibit both heat and ovulation. The duration of such a treatment must be enough to allow all animals to complete their own luteal phase and leave them all under the influence of the injections so that when these cease they will all, in a synchronous fashion, develop a follicle, come into heat and ovulate. Since the luteal phase may be over 18 days, 18–20 day treatments were used to achieve this effect.

Other synthetic compounds with similar actions to progesterone (the 'progestational agents') were also used. Some of these were fed to cattle for the prescribed period of time and proved more successful than progesterone itself. Unfortunately, however, although all these long-term progestational treatments were successful in synchronizing heat with reasonable precision the fertility of the synchronized heat was consistently depressed when compared with that of spontaneous heats in similar control animals. When shorter treatment periods were used (e.g. 12–14 days) fertility was normal but a proportion of animals failed to synchronize, i.e. those in the first few days of the cycle at the time of commencement of the treatment.

Luteolytic agents

Small doses of oestrogens, in particular oestradiol benzoate and oestradiol valerate, have been shown to cause regression of the corpus luteum and inhibition of its progesterone production (luteolysis) but their action has been somewhat unpredictable. With the advent of the prostaglandins, particularly those closely related chemically to the natural prostaglandin $PGF_2\alpha$, there was available an extremely consistent and predictable method of inducing luteolysis.

A single intramuscular dose of prostaglandin given to any bovine with a functional corpus luteum causes an abrupt cessation of the secretion of progesterone. This is followed, as already described, by the maturation of a follicle, the appearance of heat, and finally by ovulation. Unfortunately, those animals in the first 4–5 days of the cycle at the time of treatment are unaffected by the prostaglandin, as are those at the end of the cycle which no longer have a functional corpus luteum (see fig. 4.1). A single dose of prostaglandin given to a randomly cycling group of cattle will therefore synchronise heat in only a proportion of them (about 70%). However, if a further dose of prostaglandin is given 10 or 11 days later then all the animals will be in the sensitive middle phase B of the cycle (see fig. 4.1). Those in phase A at the first treatment will have passed automatically into phase B while those in phase B will undergo luteolysis, come into heat and ovulate, and will be back in phase B 10–11 days later. Those in phase C will come into heat and ovulate and also be in the sensitive phase B at the time of the second treatment.

Phase A 4–5 days	Phase B 10–12 days	Phase C 4–5 days
Insensitive early luteal phase	Sensitive luteal phase	Insensitive follicular phase when spontaneous luteolysis has already occurred

Fig. 4.1 Times in the oestrous cycle when the cow is sensitive to an injection of prostaglandin

This 'double spaced injection' regime of prostaglandin treatment is now a well established method of synchronizing heat in cattle and can be followed by two inseminations at the predetermined times of 72 and 96 hours after the second injection. Such a course of injection and insemination has been shown to result in normal fertility and obviates the need for heat detection, but it is by no means the only way in which prostaglandins can be exploited.

Combinations of treatment

As already mentioned, normal fertility can only be achieved with progestational treatments at the expense of good (closely timed) synchronisation. However, if the luteal phase can be shortened then there is no longer any need for the treatments to be so sustained. This shortening can be achieved with oestrogens, and the combination of oestrogen and progesterone forms the basis of one commercially available method of controlling heat, known as

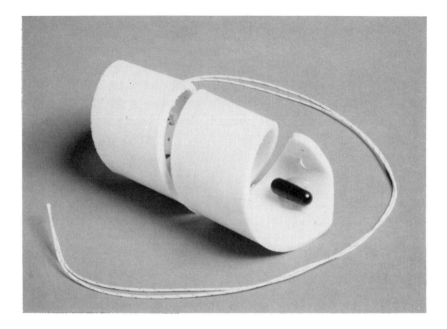

Fig. 4.2 The progesterone releasing intra-uterine device (PRID).

the 'PRID' or 'progesterone releasing intra-uterine device' (fig. 4.2).

This product consists of a stainless steel coil coated with silastic which is in turn impregnated with progesterone. Attached to the coil is a small capsule of oestradiol valerate, and the whole is inserted into the cow's vagina where it remains for 12 days. The retention rate of these devices is good. Following removal, oestrus occurs 48–72 hours later and insemination at that heat is followed by normal fertility. Alternatively, cattle can be inseminated once on a fixed time basis at 56 hours after removal or twice, at 48 hours and 72 hours, which should result in slightly better fertility.

Similar regimes of treatment using a subcutaneous silastic implant impregnated with a synthetic progestogen have also been developed; the very potent progestational compound 'norgestamet' being used in some parts of the world.

Prostaglandins have also been used to allow shortening of progestational treatments; injected either at insertion or at removal of the coil or implant. A particularly effective method is a 7 day insertion of either the PRID or norgestamet implant with an injection of prostaglandin 24 hours before withdrawal. This technique gives excellent synchronization and good fertility. There are sound physiological reasons why it should be the most effective method; in particular because there is no chance of there being any progesterone available from the cow herself which might otherwise interfere with good synchronization. However, this method is expensive to use in the field and has not yet been applied on any scale.

Advantages and disadvantages of different techniques

Progestational agents

In the U.K. at least, and probably in many other parts of the world also, it has now become accepted that the natural hormone progesterone is the most satisfactory progestational agent to use and that the stainless steel spiral is the best method of administering it.

As already pointed out, the retention rate of these devices is high (above 95%) but they are not easy to insert quickly and hygienically and inevitably, after 12 days in place, they do cause a degree of inflammation of the vagina and an evil smelling vaginal discharge is expressed when the coil is removed. This latter problem does not interfere with fertility and herdsmen can be reassured on this point.

The coil is difficult or impossible to use in maiden heifers, but for ovulation control in adult beef and dairy cows it is now a well established technique. Its main advantage over prostaglandin use is the substantial body of evidence now accumulating which shows that the PRID can be capable of initiating cyclicity in post partum non-cycling beef and dairy cows, or at least in those which are close to cyclicity. It may therefore be used in the early post partum period to breed animals which are being 'pushed forward' in the breeding season and may be more effective in some groups of lactating cattle, particularly where a proportion of them are likely to be either not cycling or cycling irregularly. The PRID does not cause abortion in pregnant animals and it is therefore not essential to examine all cows before treatment, although many people would claim that rectal palpation is an essential prerequisite of any of these treatments. Although the treatment does not cause abortion, it is essential to remember that if fixed time insemination is being used then the inseminations themselves will interrupt a pregnancy.

The prostaglandins

The prostaglandins are very easy and economical to administer. They are also very precise and predictable in their physiological effect. The double injection/double insemination regime is well proven; it gives very good results in maiden heifers, but somewhat poorer results in cows.

Since prostaglandin injections can result in the abortion of pregnant animals, rectal palpation prior to treatment is of the utmost importance. The main problem associated with prostaglandin application is that of non-cycling or irregularly cycling cows. Careful veterinary screening prior to treatment helps to minimize the problem, but indiscriminate use of prostaglandins (PG) can lead to expensive disappointments.

The most important advantage of prostaglandins over any other available technique is their versatility, for although the double injection/double A.I. method was the first to become well established there are now almost as many regimes of treatment as there are managemental situations. As can be seen from the following list, they involve a greater or lesser degree of management and veterinary intervention.

1. 2 PGs; 11 day interval; 2 fixed time A.I.s.
2. Rectal examination, 1 PG; 2 fixed time A.I.s.
3. Rectal examination, 1 PG; one A.I. on heat detection.
4. Rectal examination, 1 PG; heat detection and A.I.; 2nd PG at 11 days for cows not served and 2 fixed time A.I.s.
5. 5 days heat detection and A.I.; 1 PG for cows not served; 5–7 days heat detection and A.I. for all other cows.

Regime 1 is widely used for maiden heifers where heat detection is difficult or impossible. Regimes 2 and 3 are applicable when veterinary examination has confirmed the presence of an active corpus luteum. The most commonly used regime is 4 which is particularly applicable to dairy cows where heat detection is available but not necessarily highly efficient. All animals selected for regime 4 will be inseminated within 15 days of selection, and that is valuable assistance for a farmer. Regime 5 has been more generally used in beef cattle where holding and handling large numbers of cattle may be difficult, and also on day 1 of the breeding season in dairy herds when perhaps a considerable number of animals are available for service.

It should be emphasized again that these are only a few of the many treatment systems which have been developed in response to various managemental pressures. The selection of the appropriate regime is a matter for the veterinary surgeon in consultation with the farm staff, and might depend on such factors as:

• Numbers of cattle to be involved at one time.
• Handling facilities available.
• Labour available for handling and heat detection.
• Reproductive status of the cattle (i.e. proportion of animals thought likely to be cycling regularly).
• Costs of the exercise.
• The wishes and policies of the farmer.
• The anticipated level of fertility in the group from prior knowledge of the cattle and their nutrition.
• Understanding by the herdsman of just what is involved and the level of skill to carry out the practices.

Managemental and economic advantages of controlled breeding

In its simplest terms, the control of the bovine oestrous cycle and of ovulation is merely a method of exploiting the use of artificial insemination and, therefore, of high quality genetic material. However, the practical and economic implications of its use go considerably further. It may seem ironical now that, although the synchronization of heat cycles was a well established research objective for many years, little thought was given to which classes of stock might benefit most and what managemental advantages might follow.

Beef cattle

Beef suckler cows represent the largest and most obvious single objective for planned breeding. Most of them are kept in numbers and circumstances which preclude the use of A.I., while the cheap use of high quality selected genetic material could give substantial returns. In spite of this, beef suckler herds have not as yet been a fruitful area for the sale of controlled breeding drugs. There may be many reasons for this, among them the following:

(a) General management levels of suckler beef 'cow and calf' operations, both in the U.K. and overseas under range conditions, are very low. Nutrition is often poor and fertility is therefore also often poor. The harvest of calves of increased value born from A.I. bulls is too low to justify the expense and trouble involved.
(b) There is an inadequate supply of high quality beef semen available. Breeders often consider their own bulls to be superior to those available for A.I., and they can be right! There is also some resistance from breeders who wish to maintain a 'scarcity value' for their pedigree stock.

There should be some managemental advantages to controlled breeding in beef herds:

(a) Tightening up of the calving pattern, moving the pattern to early spring or early autumn.
(b) Batch rearing and management of groups of calves of similar age and weight.
(c) Better and more predictable nutritional management of the breeding cows.

The background level of management in beef herds is generally too low to make exploitation of such possibilities relevant.

Dairy heifers

Controlled breeding of dairy heifers can lead to:

(a) Better control of their entry into the herd.
(b) A tight calving pattern.
(c) The opportunity to use A.I. and therefore to select easy calving A.I. bulls.
(d) Most importantly, the speeding up of the rate of genetic improvement in the herd by the use of dairy semen on pure bred heifers (which should be the best genetic material in the herd). This also increases the number of heifers reared, which is often the limiting factor to herd replacement management and to herd expansion.

Heifers in particular respond consistently well to prostaglandin treatments and are therefore good candidates for heat synchronization. However, although planned breeding programmes are still carried out in dairy heifers demand has fallen, probably for the following reasons:

(a) Farmers think most of the managemental advantages can be achieved by good management and careful use of a bull.
(b) Increasing veterinary and, more particularly, insemination costs.
(c) The growing realization that, after going to a considerable amount of trouble and expense, the number of heifer calves which result can be very low even when the programme itself has been perfectly successful.

Perhaps the most important potential application of controlled breeding in dairy heifers is the production of heifers to calve at two years of age in herds with a tight calving pattern. As already mentioned (chapters 1 and 2), it is generally considered preferable to calve heifers early in the calving season in order to give them a little longer to get back in calf during lactation. It is very difficult to arrange to have an adequate supply of heifers ready to calve at two years at this time. The best source of such heifers are the maiden heifers themselves, and in order to arrange for them to grow as a group at the correct speed and to be inseminated at the correct time a controlled breeding programme is essential.

Dairy cows

Controlled breeding, and particularly prostaglandin use, comes into its own in the dairy cow. We have already emphasized several times the importance of maintaining good individual cow calving intervals for efficient and profitable milk production. The most important component of this is making sure that the cow is served at the correct time after calving.

Conception rate can be altered to some extent by careful feeding, but in a well managed herd dramatic alterations are unlikely. Heat detection efficiency, however, can be raised to 100% by strategic use of prostaglandins at the appropriate time after calving. The various regimes of treatment are described earlier in this chapter.

It must be obvious by now that in the absence of very early (i.e.

in the first fourteen days after service), accurate methods of pregnancy diagnosis any controlled breeding programme must be followed by some form of heat detection and re-breeding to pick up animals which have failed to conceive first time. In the beef herd, and often in dairy heifers, these two functions might well be performed by a 'sweeper' bull or two running with the herd. This is rarely the case with dairy cows and it is therefore fair to say that controlled breeding in general, and prostaglandins in particular, only give control over the timing of the first service after calving.

Synchronisation of an entire dairy herd is unusual, unnecessary, and may even be undesirable. The extent to which prostaglandins might be used to ensure correct timing of the first service is governed by a number of factors:

(a) Pressure on heat detection. Use may be high in herds managed by one man who likes his time off or who is busy with other farm jobs. Use often becomes high in herds where the herdsman has decided to leave and has lost his sense of purpose, and it is frequently high in herds which calve all the year round making heat detection a constant problem.

(b) Herds with a tight calving pattern often have very efficient heat detection since it is only necessary for a very short period of time. However, at the beginning of the breeding season there is always a group of cows which have calved early and are always overdue for insemination. There is a sound economic case for bringing these into heat on the first day of the breeding season.

(c) Seasonal pressure of work, e.g. silage-making and harvest, can have a significant influence on controlled breeding use.

(d) Batches of cows may be presented for prostaglandin treatment to ease the burden of heat detection in large herds. These may be identified by the farmer as batches that are losing time, or they may be selected early in an attempt to advance their calving.

Whatever the factors that stimulate the demand for controlled breeding, there is no doubt that it works best when the veterinary surgeon is making regular routine visits to the herd – as described in chapter 6. He is then in a position to select the appropriate animals and regimes of treatment and, perhaps most important of all, to maintain continuity of the breeding programme.

It should also be emphasized once again that the use of controlled breeding, whether with prostaglandins or the PRID, only influences the timing of the first insemination after calving.

Continuous monitoring of performance is essential in order to exploit fully the primary interference.

Factors influencing success

Wherever ovulation control methods are used, be it in a large or a small number of animals and whatever the class of stock, there are a number of important prerequisites for a successful outcome:

1. The nutritional management of the cattle must be good. Heifers should be the correct weight and, if possible, on a rising plane of nutrition. The complex balance of input and output in the dairy cow must be correct and body condition should be adequate. Similar constraints apply to the beef cow. Inadequate or inappropriate nutrition will lead to low conception rates and disappointing results.

2. General management during the controlled breeding period is also important, particularly with regard to the minimization of stress. It is much better, for example, if the gathering of cattle for treatments and inseminations is not used as an opportunity to worm or vaccinate or freeze brand the same animals. Changes of environment are particularly to be avoided. In the bovine changes of this kind can represent very considerable physiological stress and such stresses are known to be capable of interfering with the animal's responses to hormones, particularly the ovulatory response, and also of depressing conception rate.

3. Good handling facilities are also necessary for the minimization of stress, and preferably these should be facilities to which the cattle are already accustomed.

4. Herd records and cattle identification must be good. Some regimes of treatment are complicated and following large numbers of cattle through them is not always easy. Many mistakes have occurred through inadequate identification and recording.

5. Associated with this, and particularly important in the beef herd, is the confirmation that the correct post partum interval has elapsed before controlled breeding is attempted.

6. It must not be forgotten that good quality semen is essential for good results and that this semen should be handled properly by well trained inseminators working under conditions which are conducive to easy, stress-free inseminating.

The general message must then be that controlled breeding,

whether of one animal or a number, needs careful planning of all its aspects. Full discussions between farmer, veterinarian and stockman are essential, well ahead of time, to ensure that all the details of the exercise have been considered.

5 Fertility management

On most dairy farms fertility is a managemental rather than a physiological problem. To a great extent the fertility status of his cattle is entirely within the farmer's own control, and there is much more to fertility management than simply phoning for the inseminator to come. Similarly, it is not merely a question of achieving a good conception rate or a good calving index. The areas to be covered in this chapter therefore include:

- Consideration of the term 'fertility factor' and how this can affect herd performance.
- The policy of the farmer as it affects the interval to first service.
- The use of A.I. and natural service, and the advantages and disadvantages of each.
- The problem of heat detection and some of the aids which can be used.
- The maintenance of good pregnancy rates.
- The regularity with which cows are served repeatedly.
- The culling policy for cows failing to conceive.

Many of these factors are interrelated and of course impinge on other factors as well. They are also all related to the fertility factor of the herd itself. Understanding these fairly simple concepts can save an enormous amount of money in modern dairying.

Fertility factor

The main variables affecting calving intervals and culling rate are heat detection (HD) and pregnancy/conception rate (CR). If these are considered together, then the product of the two can be called the fertility factory (FF).

$$HD \times CR = FF$$

An average FF might therefore be:

$$60\% \times 50\% = 30$$

Another way of considering the fertility factor is to estimate how many cows in any herd might become pregnant during a 21 day period (one cycle length) of heat detection and insemination. If 100 cows in a herd of average performance were subjected to such a programme then, as seen above, only 30 pregnancies would result. Most farmers' estimates of this figure would be much higher.

Table 5.1 Effect of fertility factor and length of breeding season allowed on calving index and percentage of cows failing to conceive

Culling after oestrous cycles (no.)	Fertility factor			
	20		50	
	Calving index (days)	% failing to conceive	Calving index (days)	% failing to conceive
5	378	33	360	3
7	390	21	365	1
9	400	13	366	—
∞	429	—	366	—

Note: No animals served before 50 days post partum

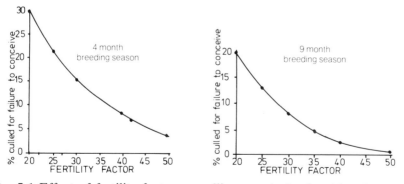

Fig. 5.1 Effect of fertility factor on culling rate in herds with calving spread of four months (LHS) or nine months (RHS)

Since this is a mathematical concept, the effect of fertility factor on both calving to conception intervals and culling rates after a nominated number of heats can be worked out. These relationships are shown in table 5.1 and fig. 5.1, and it is apparent that a high level of fertility factor permits the target calving index to be achieved without sacrificing the level of culling for failure to conceive, and *vice versa*. As the fertility factor decreases this flexibility is drastically reduced, with a good calving index only being achieved at the expense of a high proportion failing to get in calf.

In a closed herd it is also possible to use this concept to illustrate the importance of producing heifer calves in the first two months of the calving season (see fig. 5.2 and table 5.2).

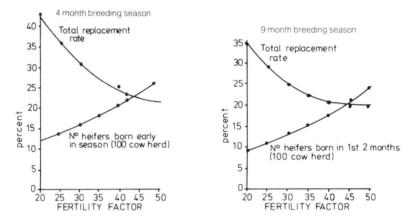

Fig. 5.2 Effect of fertility factor on heifers born in first two months of calving season (LHS calving spread four months; RHS calving spread nine months)

Table 5.2 Number of heifers born in first two months of calving season at Fertility Factor 30 (100 cow herd)

	Total replacement requirements	Cows only	Heifers out of cows and heifers
9 month breeding season	24	13	22
4 month breeding season	31	16	27

In both short and long calving seasons it is only possible to produce sufficient heifer calves to meet requirements when the fertility factor is 45 or above. If this is not achieved the farmer has the more expensive options of:

(a) Purchasing additional heifers.
(b) Allowing later born heifers to be reared for calving down at 2 years 6 months to 2 years 9 months.
(c) Allowing the later born heifers to calve at 2 years of age but later in the calving season (and hence in the less profitable months) and extending the calving season in the first and subsequent years. Alternatively, he can breed heifers out of heifers. If heifer calves are retained out of early calving heifers bred to dairy sires, then the number of early born calves can be increased significantly.

Unless heifers are bred out of heifers the supply of replacements cannot meet demand. Even when they are, a fertility factor as low as 30 means that supplies do not quite reach the necessary level. Such a fertility factor represents the typical U.K. achievement.

In herds with 9 month breeding seasons, the effects of the fertility factor on calving interval show that as it increases from 30 to 50 the mean calving to conception interval decreases by 20 days (from 393 to 373 days – fig. 5.3). The wider spread of individual calving to conception intervals in the herd with a fertility factor of 30 as opposed to 50 shows why extra culling takes place in the former herd.

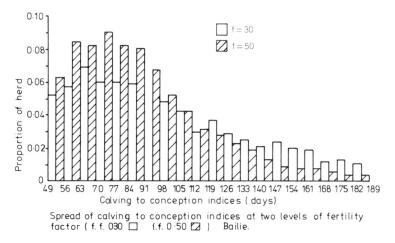

Spread of calving to conception indices at two levels of fertility factor (f. f. 0·30 ☐   f.f. 0·50 ▨)   Bailie.

Fig. 5.3 Spread of calving to conception indices at two levels of fertility factor

As many as 49% of the calving to conception indices fall between the target 49–97 days when the fertility factor is 50 (as opposed to 40% when it is 30).

Farmer policy

Calving patterns

Choosing the calving pattern depends on the objectives of the business. These patterns will of course directly affect the choice of the months of the service season. Commonly the serving season runs from early November or early December through to May, June or July in autumn calving herds, and from late March or early April until June or July in spring calving herds. The most profitable months to choose are described in chapter 2. In other herds there are no specific serving seasons.

Interval to first allowable service

The interval to first allowable service from calving (the 'serve-after date') should generally be not less than 45 days. Cows giving more than 40 litres and first lactation heifers giving more than 30 litres should not be served until 50 days post partum. It may be better to leave first lactation heifers until 50 days in any case, and the same may apply to cows which have had difficulty at calving.

Generally pregnancy rates are affected by the interval post calving at service (table 5.3) but there are circumstances for forgoing the highest pregnancy rates in order to achieve ideal calving intervals. Cows calving late in the season have, in a seasonally calving herd, very little time to be got back in calf. For instance, in a well managed autumn calving herd calving from 1 September to 30 November there will be a serving season from 1 December to 28 February. This means the animal calving on the 1 September has three months of serving season to become pregnant, none of which occurs at less than 90 days after calving. This has two effects. Many of the animals calving early in the season get in calf between 100 days and 130 days after calving, and hence calve a month later in the subsequent year than the current year. This means that, unless heifers are introduced into the herd very early in the season in sufficient numbers, a slight drift occurs in calving pattern; or if not a drift then the animals calving in the

Table 5.3 Effect of interval to first service on interval to conception and other fertility indices in one herd in four seasons

						Days to first service								
	0–35	36–41	42–48	49–55	56–62	63–69	70–76	77–83	84–90	91–97	98–104	105–111	>111	Average
No. served	36	49	145	187	155	145	128	108	67	47	39	19	71	1196
Pregnancy rate to first service %	13	34	36	39	42	40	38	43	41	55	53	52	54	41
Calving to conception (days)	90	89	92	89	93	104	108	122	123	112	124	132	153	105
% conceiving eventually	77	85	90	91	92	87	96	96	92	93	94	89	83	90
First service – conception (days)	65	51	47	37	34	38	35	42	36	22	23	24	7	36
Serves/conception	2.7	2.4	2.5	2.4	2.2	2.3	2.4	2.5	2.3	1.8	1.9	1.8	1.6	2.3
Lactation length (days)	297	297	289	288	295	307	315	320	309	323	345	326	343	304

Source: Daisy – The Dairy Information System, University of Reading

first two months of the calving season swop their calving months with each other from year to year. The early calved cows are also likely to be culled for failing to conceive even before the serving season ends because they are bound to be late to first service and they reach the 'culling decision stage' (a rather vague stage on many farms, often somewhere between five and six months after calving) when extra efforts to get the cow in calf are considered wasteful since the animal will be dry for so long. On average every extra day on the calving to conception interval only leads to two-thirds of a day on the lactation length (table 5.3).

Another problem of seasonally calved herds is that many of the late calved cows either get inseminated so early (less than 35 days after calving) that their calving to conception intervals actually lengthen, or they fail to conceive by the end of the limited serving season and hence a large proportion of the cows calving in the last month of the calving pattern are culled wastefully.

The use of A.I. or natural service

Because of apparently better detection rates many farmers, particularly later on in the serving season, use natural service rather than A.I. However, the risks of very early conception, danger to staff and the bull's own infertility can be successfully avoided by the well organised use of A.I. In the case of small herds it is expensive to keep a bull for relatively small amounts of work. In addition, of course, artificial insemination allows much safer and faster genetic progress to be achieved.

For most people the choice centres on whether natural service would overcome problems of heat detection on the farm. When replacements are either not needed from late calving cows or sufficient are thought to already be in embryo then the use of a beef bull to 'sweep up' is now common practice.

Where natural service is used the herd usually needs grouping so that the bull is only with cows that are fit to serve (i.e. are 45 days post calving and are not to be sold empty). Often a bull will get out of his group and serve cows that are really too soon after calving or serve maiden heifers when they are too small, too young, or in close season – or indeed when they are from the next door farm! The whole technique often gets out of control because, far from being the easy option, it is in many cases more difficult to manage a bull than to deal with A.I. and heat detection. Many farmers

Table 5.4 Cost of keeping a bull for 12 months

	£
Depreciation £100/year (£1000–£600 in 4 years)	100
Feed 3 kg concs/day at £150/t	164
Veterinary and medicine	30
Grazing variable costs	72
Bedding ($\frac{1}{2}$ tonne straw)	13
Fixed costs (housing, service crate, share of fixed costs)	400
Interest £1000 @ 10%	100
	879

Cost per service, say, 160 in 100 cow herd = £5.50

underestimate the cost of keeping a bull (see table 5.4).

Artificial insemination was originally developed in order to reduce costs in small herds where keeping a bull was uneconomic and also to prevent the spread of venereal disease. The added advantages of safety for the herdsman was also a significant feature. These considerations are still relevant.

Heat detection

Oestrous behaviour

Where A.I. is used, cows must be correctly detected in heat to be submitted for insemination. Even when a bull is used in a bull pen, heat detection is essential. Assuming that 95% of cows return to cyclicity within 42 days of calving, then the problem is one of the detection and not the exhibition of heat. Cattle show heat regularly, with 95% of returns to heat being within 16–28 days of the previous heat. About 90–95% of cows cycle regularly in normal herds, the remaining cows being anoestrus if only temporarily.

It is a fact that average heat detection rates by herdsmen is 55% even when cows do exhibit behavioural signs. Every effort must be made to raise heat detection rates to 80% of cows that are fit to be served, to be detected in a 24 day period – without, of course, sacrificing the accuracy and quality of detection.

What to look for

There is a 90% certainty that the cow that *stands* solidly to be mounted without walking away is in heat. There is only a 30% chance that the cow that mounts another is in heat. The key feature is, therefore, standing to be mounted.

This type of behaviour lasts only for a short time (about 8–12 hours in cows, 6–8 hours in maiden heifers). Twenty per cent of heats last for less than six hours. In addition, possibly because the detected cows are often removed from the herd to await A.I. during the day, 65% of the standing heats occur overnight (between 6 p.m. and 6 a.m.).

Apart from aggressive behaviour, a cow only really touches another when within 18 hours (either side) of heat. The general build up is shown in fig. 5.4 and involves a sequence of:

1. Licking and rubbing each other.
2. Sniffing of the vagina of another cow.
3. Mutual chin resting.
4. Lining up to mount another cow.
5. Mounting another cow.
6. Standing to be mounted by another cow.

Items 1–5 then occur in reverse order.

This is only a general pattern and not a sequence to be expected in individual cows. The most certain behavioural signs to look for are:

- Standing to be mounted.
- Mounting head to head (top cow).

Other often circumstantial evidence to look for is:

- Rubbed tail head and tail bones (rubbed raw).
- Muddy, rubbed back and sides.
- Steaming back (and areas around tail).
- Daily yield down more than 10% compared with the yield over the last seven days, followed by a rise in yield often on the best day for service.
- General restlessness, off feed, entering milking parlour out of usual order of place.
- Bloody mucus from vagina (this usually means that heat has been missed).

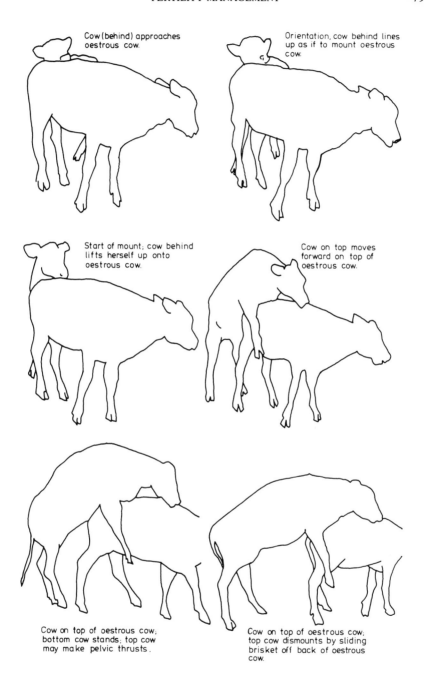

Fig. 5.4(a) Sequence of oestrous behaviour in cattle

Not standing to be mounted;
bottom cow takes evasive action
by walking or running away.

Chin resting; cow behind rests
chin on back half of recipient.
Poor sign of oestrous behaviour.

Sniffing and licking;
licking back half of recipient.

Head to head. Top
cow is in oestrus.

Licking; licking front half of
recipient Not strongly
associated with oestrus.

Fig. 5.4(b) Other less useful types of oestrous behaviour

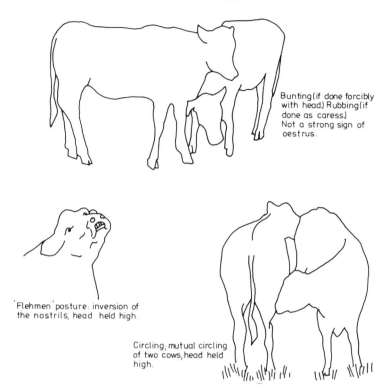

Bunting (if done forcibly with head.) Rubbing (if done as caress.) Not a strong sign of oestrus.

'Flehmen' posture: inversion of the nostrils, head held high.

Circling; mutual circling of two cows, head held high.

Fig. 5.4(c) Other less useful components of oestrous behaviour

Table 5.5 Length of heat period (two studies)

Length (hours)	Study 1	Study 2
2–6	20%	14%
6–12	50%	54%
12–18	25%	29%
18–24	4% ⎫	4%
Over 24	1% ⎭	

The time period over which oestrous behaviour is exhibited is controlled physiologically. However, the frequency of display of the various components of behaviour during this period is governed by environmental and social factors, in particular the presence or absence of another cow in oestrus. Average durations of oestrus are shown in table 5.5.

Detection rate

Recent work has shown that on-farm heat detection rates can only be raised insofar as finding the first service is concerned. Returns to service were detected at constant rates despite considerable efforts by advisers to improve this. There may, therefore, be two rates of heat detection in dairy herds. Indeed, there may also be effects of season on heat detection: firstly, as time passes the herdsman's efficiency and enthusiasm wanes; and secondly, in very cold or very hot weather the level of heat behaviour shown may be reduced. Day length may also affect heat behaviour and short days are likely to reduce its incidence.

The number of cows in heat together affects heat behaviour. The number of mounts attracted by an animal rises to a peak when three or four animals are in heat together and then declines (table 5.6). Standing to be mounted is the definitive sign, but it remains difficult to spot since a quarter of cows are mounted less than 30 times and much of the activity occurs at awkward times, e.g. between 6 p.m. and 6 a.m. These times may be partly due to the fact that the management routine of most herds is unlikely to allow the cows much chance to interact during the normal working day, busy as it is with milking, feeding, slurry scraping and bedding, etc.

Other signs of heat are less distinctive as they occur over a wider spread of time. What is being sought is a sign that shows when the cow is ready for insemination. Usually the best time to inseminate is when the cow is coming to the end of the standing to be mounted period (see chapter 3).

Breaks in standing behaviour do occur in cows on heat so herdsmen should be aware that 25% of cows may not stand at one observation in the middle of their period of behaviour. The daily pattern of heat behaviour allows the alert herdsman to raise heat detection rates to 80% and aids may help even the keenest herdsman to achieve higher rates. Every help should be given to enable cattle to show signs of heat. This means adequate planes of nutrition should be provided and slippery floors should be eliminated.

Table 5.6 Effect of number of animals in heat together on mounting and standing behaviour

	No. in heat together			
	1	2	3	5
Duration of mounting behaviour (hrs)	15	16	22	19
Total length of all oestrous behaviour (hrs)	23	32	41	26
No. of mounts (by and to the animal)	54	109	132	132

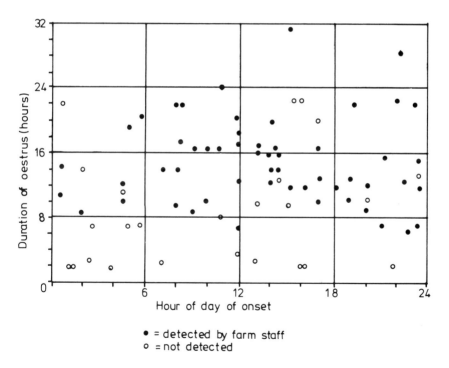

● = detected by farm staff
○ = not detected

Fig. 5.5 Effect of duration and time of onset of oestrus on whether or not detection occurred (60 cows 80 oestruses)

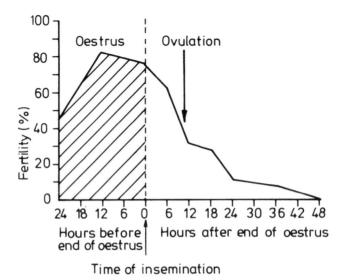

Fig. 5.6 Influence of time of insemination on conception rate in cattle. The optimum time of insemination for maximum fertility is some 12–18 hours before ovulation. Late insemination causes a drastic fall in potential fertility. (*From Hunter.*)

Heat detection techniques for the stockman

The following points emphasise some of the ways in which heat detection can be improved:

1. Identify the cows with a freeze brand and an ear tag.
2. Choose times for observation when cows are not being disturbed.
3. Use *standing to be mounted* as the best criterion of heat.
4. Use a notebook.
5. Observe three or four times per day, leaving not more than eight hours between visits. This is practically guaranteed to lift rates to 80%. (See table 5.7.)
6. Move through the herd checking cows seen in heat at previous times as well as cows freshly in heat.
7. Watch for at least 20 minutes at a time. This increases the rate to 85% as against 65% when only 15 minutes is taken.
8. Record the time the cow first comes into heat, in order to help

time the insemination at *12–18 hours* after cow first stands to be mounted.

9. Adopt a *team approach* on the farm and approach the job in a purposeful manner with good recording and analysis techniques. Giving a score of 4 to the cow seen standing to be mounted and ranging down to a score of 1 for the cow seen with lesser signs has led to better, more accurate heat detection.

Table 5.7 Effect of number and times of observation on heat detection rates

No. of observations	Observation time					Detection rate
2	0600		1800			69
2	0800		1600			54
2	0800		1800			58
2	0800		2000			65
3	0800	1400	2000			73
3	0600	1400	2200			84
4	0800	1200	1600	2200		80
4	0600	1200	1600	2000		86
4	0800	1200	1600	2000		75
5	0600	1000	1400	1800	2200	91

How to measure heat detection efficiency

A range of methods exist to measure heat detection. All of them require the recording of service dates at least. By knowing the interval between services the information can be assembled to show the percentage of interservice intervals that are within a normal range, e.g. 16–28 days. The target is for 80% of returns to be of this length.

Another method is to decide how many extra 21 day intervals fit into the interservice interval involved, e.g. a return interval of 63 days implies two intervals have been missed while a return interval of 38 days implies one has been missed. These missed oestruses are added up and used in the formula:

$$\frac{\text{Detection}}{\text{rate}} = \frac{\text{No. of interservice intervals in fact}}{\text{No. of interservice intervals + no. of missed oestruses}} \times 100$$

This calculation underestimates heat detection by about 5%, depending on the level of early embryonic mortality.

A third method is to compose the interservice intervals into the formula:

$$\frac{(\text{No. 18--24 days}) + (\text{No. 36--48 days})}{\text{All interservice intervals}} \times 100$$

This method assumes that the longer returns (36–48 days) are not a question of missed heats, and hence seriously overestimates the level of heat detection.

It has been common to use the mean interval between services as a way of expressing heat detection rates. Using this method, a mean of 42 days implies a 50% detection rate, one of 21 days 100% and 26 days 80%.

$$\text{Detection rate} = \frac{21}{\text{Mean interservice interval}} \times 100$$

This method does not take proper account of the short or long cycles which may have arisen through inaccurate heat detection, hence with a lot of short cycles it can seriously overestimate the detection rate.

One way of illustrating the importance of accuracy as well as rate of heat detection is to compose a histogram showing the interservice intervals and their rate of occurrence. The percentage of return intervals that are below 15 days should be less than 5%, the percentage between 16 and 28 days should be 75% and those between 32 and 56 days should be 10%, with only 5% of longer duration (see fig. 5.7).

Aids to heat detection

Tail paint

A well placed strip of tail paste or any water based emulsion paint is a cheap and effective aid. It becomes scuffed, removed or cracked when one cow mounts another.

The paste should be put on using a stiff 2–3 inch paintbrush, the appropriate part of the tail head having been brushed to remove loose hair and soiling. The paste should be stippled into the hair at the rubbing point, making sure it has penetrated to skin level, with

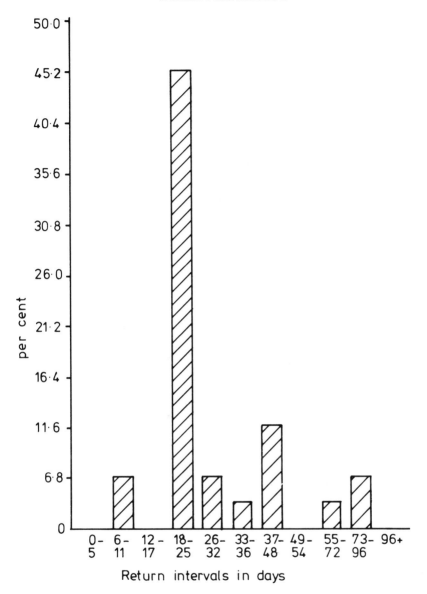

Fig. 5.7 Heat detection analysis by interval between serves

a final backward stroke producing a smooth strip which should ideally be about 8 inches long and 2–3 inches wide.

The cows should be checked at least once a day otherwise heats may be missed. In parlours which have a sunken pit, a mirror (or a piece of foil-covered plywood) fixed at an angle over the entrance improves the ease with which the herdsman can see the strips of paste.

Kamar Heat Mount Detectors

These pressure pads, consisting of an outer white tube and an inner phial of red dye which is squeezed out by the mounting cow (fig. 5.8), are placed on the tail head. The brisket of the top cow squashes the Kamar which changes colour when the cow stands to be mounted for more than a few seconds. The pads can be used in the same way as tail paste but they are about seven times more expensive. False positives do occur, for example when the cattle are closely confined or when there is rubbing of the crucial areas on such items as cubicle rails.

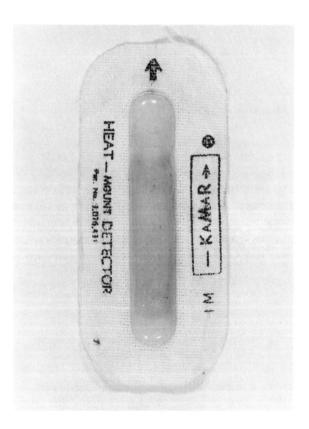

Fig. 5.8 Photograph of Kamar Heat Mount Detector

Vasectomised bulls

These have been used quite successfully with chin-ball markers. Normally they should be worked in pairs with one resting and one working in two day periods. Some skill is required in detecting the marks on a cow truly in heat and in keeping the supply of ink topped up. Penis deviation may be preferred to vasectomy, and an 'androgenized' female may have the same effect but be safer. In the latter case male hormones (testosterone) are administered over a matter of weeks to a heifer or cow that is not to be milked. The animal becomes 'bull-like' in behaviour. The treatment works on only about 7 out of 10 animals and leads to rapid weight gain in the female which necessitates its sale for being too heavy after 3–6 months or so.

Unfortunately individual animals treated in these ways behave differently and it is important for the stockman to be able to recognise these differences in order to interpret his results.

The establishment of pregnancy

A wide range of performance in pregnancy rates occurs, ranging from 20% to 75%. Performance at a lower rate than 50% can have a major effect on the economics of milk production due to waste of time, lower production, extra semen costs and high replacement rates. Even within the same herd pregnancy rates can vary considerably within the same season as a result of a wide range of factors such as the timing of insemination, the nutrition of the cow and the fertility of the semen used.

Reasons for poor pregnancy rates

Timing of insemination

As mentioned in the section on heat detection, timing is important. For best results cows should be served 12–18 hours after they first stand to be mounted. A 24 hour range must, however, be tolerated because of the once-a-day insemination service. Herdsmen looking for a general rule (and assuming the A.I. man calls at 12 noon) should put forward cows first seen on heat by 6 a.m. for insemination that day, but others seen later than that should be left until the next day. Above all, the cow must really have been in standing heat and should be served the next day as well if she continues to stand.

With the advent of do-it-yourself A.I., twice-a-day insemination routines are often practised but they are not always helpful. One cause of the failures might be that some animals, particularly those inseminated in the afternoon or evening, have been inseminated too soon in their heat period. Here the routine might be to inseminate at 8 in the morning all cows first seen in heat between 10 in the morning and 11 at night thus giving the insemination at from 9 to 22 hours of heat, and at 4 in the afternoon to inseminate all those first seen in heat between 11 in the evening and 10 in the morning. This means in the latter case that cows are being served when they have been in heat from 6 to 17 hours.

Up to 22% of cows have been shown in some surveys to be inseminated when not in heat, though in well run herds this error is probably very unusual. Poor recognition of heat signs, inadequate attention to detail and poor transfer of information are often the cause. Many cows served at the wrong time are within 1–5 days of genuinely being in heat, and the stockmen are confused by the behavioural signs and do not take the trouble to distinguish a standing heat properly.

Nutrition and milk yield

More important than timing and insemination technique, the main cause of poor pregnancy rates in cattle is inadequate nutrition. High yield may often be blamed, but only in very unusual circumstances is this justified. More usually the problem is one of underfeeding energy and/or protein although this is, of course, more likely in cows giving high yields. Sometimes very frequent feeding during the day is the only way of raising feed intakes to a high enough level to overcome the deficit caused by very high yields.

The level at which yield becomes crucial and it becomes difficult to feed the cows sufficient is when they are producing about 8000 kg milk in a 305 day lactation or 7000 kg for first lactation heifers.

At the time of service, say between 45 and 100 days after calving, these animals will be giving 40 kg and 30 kg of milk respectively each day. The energy demands (megajoules) for such a cow can be calculated as shown in table 5.8. If the dry matter intake is as high as 20 kg/day of which forage dry matter intake is assumed at early stages of lactation to be no more than 7 kg then 13 kg of dry matter remain to be fed, usually as concentrates.

Table 5.8 Energy requirements of a high yielding cow

Maintenance (per day)	63 MJ
Yield 40 × 5.5 MJ	220 MJ
	283 MJ
Contribution from wt loss	
1.0 kg/day at 33 MJ/kg	−33 MJ
	250 MJ needed from feed
7 kgs forage at 10 MJ/kg DM	70 MJ
Remaining energy deficit	180 MJ
Remaining dry matter intake	13 kgs
Energy density needed of remaining dry matter	$\frac{180}{13} = 13.8$

In many cases even a concentrate ration cannot supply such energy density and the cow, if she has the reserves, loses weight faster to supply the deficit, sometimes losing as much as 3 kg liveweight per day for 70 days or so. This type of cow, under these conditions, is hard to get in calf. In fact the cow's heat signs may also be weaker than usual and, in many cases, oestrous cycles will cease. Some workers have found that three times a day milking which can raise yields by 10–20% may lead to these conditions in more cattle in the herd unless feeding levels are kept up to standard.

The cow should be in a stable or improving condition, and generally that means gaining weight, at the time of service. It seems important to stem the tide of weight loss in recently calved cows as quickly as possible, and to stop the absolute loss from going too far. This 'trough weight' should be reached at about 35 days after calving, and from then on feeding practices should be instituted to raise liveweight. Recent work shows that cows whose yields rise quickly in early lactation from days 14 to 21 are harder to get in calf than those whose yields rise more slowly.

Other factors

There is a clear relationship between difficulties around calving (twins, calving assistance, dystokia, retained afterbirth, endo-metritis, ketosis and acute mastitis) and poor pregnancy rate. It is good management not only to try to prevent the occurrence of such conditions but also, when they do occur, to treat the cows quickly to minimise the effect. Such animals should be left up to 10 extra days to 55 days post partum before receiving their first service, and indeed should be inspected and cleared by the veterinary surgeon before they are rebred (see chapter 6).

The use of a scoring technique to describe fat cover on the animal is a crude but useful way of keeping an eye on herd and cow condition. It is certainly easier and more popular than weighing cows. The cow is scored around the tail head on a 10 point scale from 0.5 (very thin) to 5.0 (very fat). Friesian cows with a condition score below 2 at the time of insemination have poor conception rates (table 5.9).

Persistency with serving

In a breeding season there is limited time for any cow to conceive, this period being shorter for those calving late. Even the earliest calvers have a limited length of time and they also get their first service late in lactation. The usual practice is to re-inseminate cows until 180–200 days post partum or until the end of the breeding season, whichever is the sooner, and then to consider any empty cows as barren. This means that late calvers are likely to have higher culling rates, and every effort should be made to devise ways of improving heat detection rates and submission rates and to shorten the interval between services in order to get the necessary number of inseminations before the season ends. 'Repeat breeders' that arise as a class in any herd are not sick in any way, and generally repeated insemination is all that they need to get back in calf. Certainly all that many of them need is a rest before they are served again!

Table 5.9 Effect of condition score at service on pregnancy rate (three studies)

Condition score	Study 1		Study 2		Study 3 (2000 cows) % pregnant
	No. served	% pregnant	No. served	% pregnant	
< 1	9	33			
1–2	72	45	40	47	45
2–3	264	54	77	56	62
3–4	17	64	92	65	70
4+			11	82	

The target must be to serve 95% of the cows that calve and to get 95% of those served to eventually conceive. This means 90% of cows calved will conceive, and hence only 10% will have to be culled empty. Highest profits come from herds with replacement rates between 15% and 20% in total, so in these circumstances there is scope for 5–10 animals per 100 being culled for reasons other than being empty.

6 Planned animal health – the role of the veterinarian

It has become clear during recent years that the traditional relationships between the farmer and his veterinary surgeon must change. The economic implications of long and sometimes expensive treatment of individual animals are not attractive, and although the veterinarian's clinical skills remain essential to the farmer the application of them is changing. As dairy farming becomes more intensified the demand is for sound preventive medicine and proper understanding of disease and management factors which affect the production of the herd. The task of the veterinarian under such circumstances is exemplified by the part which he is able to play in monitoring the reproductive performance of the herd and relating this to other management influences. It should therefore now be quite usual for the veterinary surgeon to visit dairy farms routinely on a regular basis. This chapter is concerned with the discussion of how such visits should be organized and what benefits should accrue to the milk producer.

Very broadly, the intention of these visits is to ensure that the breeding of the herd is as efficient as is economically possible and to minimize losses of time and stock through a failure to become pregnant. This is done by making sure that cows are fit for service, that they are served at the optimum time post partum and that they are pregnant after their last service. Without such monitoring, losses may accrue through the forced culling and replacement of healthy cows and by loss of time on individual calving intervals. This point is emphasized repeatedly elsewhere in the book, and the costs have been carefully calculated. Although there are some firm guidelines to follow, which it is the intention of this section to emphasize, it should be remembered that whatever targets are set must be compatible with farm policy and with any changes in that policy which the owner chooses to make.

Before embarking on any routine fertility monitoring there must be agreement between the veterinary surgeon and the herd owner as to the detailed objectives of the scheme and the methods by which they can measure their success. Discussion should also include the possible costs of routine visiting, a critical factor being the frequency of the visits. It is tempting for the herd owner to opt for visits as infrequently as possible in order to save expense. However, it is important to remember that the longer the interval between visits, the more delay there is in generating some of the information. It particular it may become impossible to get cows served at the correct time. The number of animals to be examined at each visit may also become too large at certain times of the year, leading to problems for the herdsman both in sorting them out and in attending to the necessary paper work. The frequency of visiting should, then, be governed by the herd size, the competence of the people involved, the performance targets which are demanded and, of course, by economics.

The following are some suggested visiting frequencies:

Herd size	All-year calving visiting frequency	Seasonal/autumn calving visiting frequency
< 100	Monthly	Fortnightly in season
100–150	Monthly/every two weeks	Weekly/two weekly depending on calving pattern
150–200	Every two weeks	Weekly
> 200	Weekly	Weekly

Visiting frequency and cost are obviously closely associated, but to be more precise about cost is difficult. Each farmer will want something different from such schemes and the ability to present cows quickly and efficiently for examination and to record the necessary information varies considerably from farm to farm.

It should be emphasized at this point that the scheme described in this chapter is only one example, and individual veterinary surgeons and farmers may agree to pursue similar objectives by quite different routes.

Performance targets

The targets which should be agreed for each cow must include the following:

	Target	Range
Lactation	305 days	300–320
Pregnancy	282 days	varies somewhat with breed
Dry period	60 days	42–75
Calving interval	365 days	350–380
Calving to 1st service	not less than 50 days and not more than 60 days	
Calving to conception interval	83 days	69–98
Conception rate	60%	
Heat detection efficiency	80%	
Number of cows seen bulling and recorded as such by 60 days post partum should be	100%	
Herd calving index	350–380	
Involuntary culling	< 10%	

Pregnancy/lactation/dry period

The only unchanging and unchangeable target is obviously the length of pregnancy. The length of lactation and the dry period are included because they may well influence fertility. Ruthless drying off at the appropriate time and efficient dry cow therapy are, of course, essential components of a mastitis prevention programme. It is equally important to ensure that the cow gets adequate rest before beginning another lactation and that nutrition during this late stage of pregnancy is good. If this is not the case the cow will not be in good condition at calving and replacement of the inevitable weight loss which follows will take too long, leaving her unprepared and perhaps infertile at the time when she is expected to become pregnant once more.

Calving interval/calving to conception

The calving interval of 365 days is the basic demand of any

breeding programme, but as a measurable parameter it is of only limited interest because it is obviously historical before it can be recorded. However, it is a good exercise in any herd to calculate not only the average calving interval (calving index) but also the distribution of individual calving intervals. A good calving index (which is a commonly used figure) may be supported by a very wide variation in individual intervals and may therefore not say much about the profitability of the herd.

To understand what is happening within the herd it is much more valuable to monitor closely the events in the post partum period. All heats should be recorded whether or not they are used as opportunities for service or insemination, and a substantial number of cows seen bulling by 50–60 days after calving is a good indicator that ovarian activity in the herd is regular and normal. Cows should be served for the first time by about 65 days post partum unless there are good management reasons for not doing so, but should not be served before about 45–50 days since conception rate tends to be lower during the early post partum period. If the herd is properly managed and fed the conception rate may allow a later first service. There is no justification for serving very early after calving except to salvage some success in herds where conception rate is low.

A careful watch should be kept on the calving to first service interval, and as soon as pregnancy diagnosis is complete a note should be taken of the calving to conception interval. This provides a good up-to-date check on reproductive efficiency.

Heat detection efficiency and culling

Efficient heat detection is an essential component of good fertility whatever else may be wrong or right with the herd, and it is important to keep a check on it in any herd. Methods of detecting heat and measuring detection efficiency are dealt with elsewhere in the book, but the essential information will be generated or picked up during routine visits. It is worth noting here that a heat detection rate of 80% or more is a very difficult target to achieve. Most herds only achieve 50–60%.

Reduction in the involuntary culling of cows (i.e. cows which are sold simply because they are not in calf) is probably the most important way in which the veterinary surgeon can demonstrate to the farmer that his interference with the herd is cost effective. The target should therefore be set at a realistic level and should be properly checked from year to year. At the time of writing the cost of replacing a cow lies somewhere between £200 and £250 so the saving of, say, four cows in a 120 cow herd (i.e. around 3%) pays for a substantial amount of veterinary attention. Such a saving is not difficult to achieve.

The targets discussed here are fairly precise and may not be accepted by all farmers and all veterinarians. However, they are included because the importance of setting objectives cannot be over-emphasized. It is accepted that individual animals may have an optimal performance which is outside these figures, but present day herd size precludes the accurate management of individuals and makes some kind of average target essential. It should also be emphasized that these targets are only targets. Nobody is going to reach them all with all animals, but all animals should be measured against them.

Organisation of the visits

Having agreed the objectives and the frequency of the visits to be made there is still a choice of approach to the organization of the actual work involved. At each visit certain groups of animals will have to be examined clinically (mainly by rectal palpation) and a fully comprehensive scheme might well involve the following:

Group I All animals calved about 30 days. These cows can be examined per rectum and are sometimes also examined per vaginum or even by vaginascope. The intention is to make sure that uterine involution is complete and to identify cows which are establishing chronic uterine infections or developing ovarian cysts. 'Uterine involution' is the term used to describe the shrinkage of the uterus back to its normal size after calving. Early examination at this stage ensures that cows can be treated and be ready for insemination at the appropriate time. It also provides the opportunity to make an assessment of ovarian cyclicity. Some veterinarians may choose to do this examination earlier or later and some, in discussion with the herd owner, may decide not to do it at all. There may well be a case to be made for not examining any cow which has already been recorded as being in heat by 30 days since it is very likely to be normal. However, a fully comprehensive scheme must include some kind of pre-breeding examination and this should obviously be done in time for effective treatment to be completed before breeding begins.

Treatments selected for the conditions occurring in this group must remain the province of the veterinarian. Infected uteri will be treated by intra-uterine irrigation with antibiotics (so-called 'wash-outs') or, if they are acute, possibly with antibiotic injections. Chronic infections of the uterus, for example pyometra, may be treated with prostaglandin, which is also used to produce resolution of thick walled 'luteal cysts'. Other types of ovarian cyst may be ruptured manually and/or treated with GnRH to induce LH release in the cow and cause the cyst to ovulate (see chapter 3).

Group II Any cows calved 60–65 days and not yet served or inseminated. Such animals are obviously likely to be candidates for prostaglandin or progesterone treatment regimes and are important since successful treatment and insemination at this stage will result in the maintenance of good calving to conception intervals. Accurate diagnosis is also important. If group I animals have been dealt with successfully then group II have either not been seen in heat or alternatively are anoestrous and have inadequate ovarian function. The number or proportion of cows appearing in this group must therefore provide important information about both the management of the herd and its reproductive status.

It should be an objective of both the farmer and the veterinarian to keep the number of animals in group II to a minimum. For example, if one third of the herd was presented for examination in group II this would be considered unacceptably high. The target figure is no animals at all, but this is unlikely to be achieved if only because of the random distribution of heat dates in relation to routine visits.

The various treatment regimes for animals not seen in heat using either PRID or prostaglandin are described in detail in chapter 4. It is up to the veterinary surgeon, in discussion with the herdsman, to select which regime is appropriate.

Group III Cows presented for pregnancy diagnosis by rectal palpation. Veterinarians vary in their preferences for examining at different stages of pregnancy, but in general terms it must be said that the earlier it is done the more beneficial it is to the herd owner. It is not unrealistic to make these examinations at a minimum of 35 days after service. Cows found to be empty are either treated as though they were in group II or, alternatively, a decision may be made to cull them if they have already lost too much time. Pregnancy diagnosis later in pregnancy often leaves no alternative but to cull.

Group III animals, like those in group II, provide an important check on the status of the herd and its management. If animals presented for pregnancy diagnosis are nearly all pregnant it is safe to assume that heat detection is good and that the returns to service are being detected and used. If a substantial proportion (more than one sixth) are persistently found to be empty then it is worthwhile to examine the efficiency of the methods of heat detection being used. However, the fact that most animals are pregnant does not necessarily mean that the conception rate is good since they may all have been served several times. It is therefore a valuable exercise at each visit to note the proportion of group III cows found to be pregnant and to calculate the number of serves required for each conception. This provides a rough running check on the conception rate of the herd and the efficiency of heat detection, and also early warning when these are going wrong.

Group IV These cows constitute the so-called 'repeat breeders', i.e. any cows served three times or more but still apparently empty. If this group is not included and attention is therefore not focussed on it, the number may build up without them being presented in either group II or group III. A proportion of group IV animals is to be expected even in well managed herds of normal fertility but this should not rise above about 3–5%.

Group V It should not be forgotten that the vet's regular visit, particularly if it is on a weekly basis, provides an opportunity to examine a variety of individual clinical cases. Some may have been brought forward from previous visits (e.g. cows with vulval discharges, etc.) while others may have occurred since (e.g. cows with retained cleansings or those which have suddenly dropped in yield). This aspect of the routine visit is referred to in more detail later.

Group V should also include any cows returning to heat after having already been diagnosed as in calf. These cows must always be re-examined before being inseminated (see chapter 3) but they may be served by a bull without endangering their pregnancy.

One of the most important early benefits of this kind of scheme is that it focusses the attention of the herdsman and herd owner upon specific groups of cows within the confusion of the entire herd. The fact that the vet is coming regularly imposes a certain discipline and organization which might otherwise be lacking. The benefit is obvious in all herds but particularly in those which calve all the year round and where the management of the herd is subject to constant interruptions from other important farming pressures (e.g. harvest, silage making, etc.). It is less obvious in herds with a tight calving pattern where heat detection and insemination are only seasonal problems.

Factors affecting success

Correct cow selection is an important aspect of a successful scheme and should be checked by the veterinary surgeon involved. Clear identification of cattle and efficient methods of recording and retrieving the day-by-day events of the breeding cycle are essential.

Recording techniques are discussed in chapter 7; it is sufficient at this point to emphasize the necessity for them and to state that, in broad terms, the best technique for each farm is the one that suits the farmer and his staff, which they all clearly understand and which provides the information required.

Economics

It would be foolish to suggest that this kind of work can be done cheaply by the veterinarian. It is up to the herd owner to ensure that he has a good idea of what it will cost him, and up to him and the vet to make sure that the work is cost effective. Most veterinarians will be happy to charge for this work on an hourly basis so that the best use can be made of their time by efficient separation and presentation of the animals involved.

Regular visits also provide an opportunity for discussion and examination of routine clinical cases and general health problems on the farm and this results in a sharp reduction in the number of 'fire brigade' visits. The veterinarian may also feel that because he is closely involved in the herd management he is in a better position to dispense the necessary antibiotics and other preparations a little more freely than he might otherwise do. So although annual veterinary bills may increase the farmer should feel that he is getting something substantial in return for his money.

Other areas of preventive medicine on the dairy farm

The establishment and supervision of good clean calving facilities and the prompt and efficient treatment of conditions associated with calving, such as milk fever and retained placenta may also be considered areas of veterinary responsibility. They are certainly likely to have a bearing on reproductive efficiency. Perhaps more important still is the increasingly prevalent problem of laminitis, leading to various types and severity of lameness. This causes loss of milk and loss of condition and interferes with good reproductive performance and, as such, must be included in the general veterinary area of preventive medicine for investigation and discussion at the routine visit.

Most herd owners would probably agree that the main health problems in a modern intensive dairy herd are:

1. Fertility and the efficient operation of all its components.
2. Mastitis and the control, prevention and treatment thereof.
3. Lameness – with assorted foot problems.
4. Calf rearing – the provision of good facilities and the prevention and treatment of neonatal disease.
5. Problems at and around calving – including the establishment and maintenance of hygienic calving facilities.

All these are complex topics in their own right and demand detailed analysis and discussion beyond the scope of this book. But they are all topics which should be monitored, investigated and discussed at routine regular veterinary visits and they are all problems which can respond to a preventive approach.

Overlying all these and profoundly and inextricably linked with most of them is the problem of correct nutrition, which is also dealt with elsewhere. It cannot be emphasized too strongly at this point that reproductive efficiency in the dairy cow is much more closely dependent upon proper feeding than upon pathology and disease. If the balance between the demands of the cow for production and the maintenance of body condition are not met by nutritional intake then her fertility will fail. Cows under what might be termed 'nutritional stress' quickly become anoestrous and either fail to cycle at all or cycle irregularly. When they do show heat and are inseminated they have a much smaller chance of establishing and maintaining a pregnancy. Under such conditions the most important benefit of regular herd monitoring is early recognition of the problem, giving an opportunity to adjust the ration before too much is lost. This kind of benefit is impossible to quantify and may seem remote when everything is running smoothly.

Responsibilities of the herd owner

It is perhaps obvious by now that schemes like the one outlined in this chapter are of no interest to herd owners or managers who consider their management to be already perfect. They are intended for those who feel they need help, and it may well be that only selected parts of the routine will be of interest. It might be suggested, for example, that the essential minimum is regular pregnancy diagnosis; and if this is done regularly and early then it may be adequate. It is certainly up to the herd owner to decide what he requires and to discuss his needs with his veterinary surgeon. Before or during these discussions some important points should be considered:

1. It is necessary to have a realistic understanding of what can and cannot be achieved in the herd.
2. It is also necessary to be prepared to take a long-term view of the possible benefits. Only where the situation is bad at the outset can very rapid improvements be made. Generally speaking, improvements in productivity and profitability are gradual and when they are achieved the problem may lie in maintaining them.
3. The herd owner must be prepared to make any management changes which are necessary to get the best results. This may include the institution of good recording systems and the willingness to maintain these to ensure proper selection of cows. It may even include the provision of good facilities for the collection and presentation of cattle and the efficient use of veterinary time.

So it might be said that these schemes are not for the very well managed herd, nor for the very badly managed herd which will never realize the proper benefits. They are for the majority of herds, which lie between these two extremes.

Measuring the benefits

If the decision is made to institute very sophisticated recording, retrieval and analysis systems such as computer programs (see chapter 7) then calculation of the success of fertility monitoring will look after itself since it will be an integral part of the program. For most herds, however, something much simpler will suffice. Figure 7.3 shows the kind of chart which can easily be prepared for any herd. Cows calving during any month are listed, their subsequent breeding history is recorded and the calving to conception interval is measured. Problem cows are highlighted and those which are to be culled are clearly marked. A more complete record can be prepared annually and all these are then available for regular examination and comparison with their predecessors.

7 Records and their analysis

Keeping accurate records on a farm is essential for two main reasons. Firstly, it is the only sensible basis for taking decisions about alterations in farm policy such as changes in stocking rate or the introduction of new enterprises. Secondly, it is obviously the only way of ascertaining that the targets which have been set for production and reproductive performance are being achieved.

Target areas in which improvements might be made would include:

- Calving pattern.
- Calving to conception interval.
- Pregnancy or conception rate.
- Calving to first service.
- Culling rate.
- Incidence of mastitis and lameness.
- Calf mortality

Information comes from records through their proper collection and analysis – a process which, in itself, stimulates farm staff to keep better records.

Records to keep

The selected system should include the following elements:

1. An individual record card for each animal on which events in the animal's life can be listed in chronological order.
2. A rotary board or other form of total herd wall record.
3. A list of cows in calving date order with service dates, P.D. dates, and calving to conception interval.
4. A diary or notebook in which the herdsman records daily events occurring in the herd.

Individual cards

The information stored on individual cards would include:

(a) Name/identity (ID) – unique to the herd, clearly understandable; e.g. 123A, 192 (branded on the cow and clearly seen; large ear tag as well if possible).
(b) Date of birth.
(c) Lactation number.
(d) Calving date for each lactation.
(e) Problems at calving, abnormalities, abortion, difficulty at calving (dystocia), calf alive/dead, single or twins, breed of calf, sex of calf, identity of calf, fate of calf (died, sold, kept).
(f) Retention of afterbirth (RFM for 'retained foetal membranes').
(g) Any vulval discharges or whites (VLD).
(h) Dates of all heat periods observed without service ('bulling not served' or BNS).
(i) Service dates and the bull used, inseminator, type of heat behaviour – standing to be mounted (STBM), mounting another cow, not seen standing, raw tailhead, etc.
(j) Pregnancy diagnosis (PD) type (milk progesterone, doppler probe per rectum by the veterinary surgeon), date and result including estimated date of service if no record of service exists.
(k) Veterinary treatments and dates
    (i) reason cow put forward for vet, e.g. oestrus not observed (ONO), or no-visible oestrus (NVO), failure to conceive (FTC), irregular cycles (NYM for 'nymphomaniac'), mastitis, lameness, digestive upset, injury, endometritis (ENDO), vulval discharge (VLD).
    (ii) Finding or diagnosis in reasonable detail, e.g. corpus luteum in left ovary (CLLO), cystic ovary (CO), mastitis in left hind quarter (Mas. LH), etc.
    (iii) Treatment used (in stockman's terms) e.g. 'Orbenin', 'Estrumate'.
    (iv) Further action to be taken (e.g. see again in two weeks).
    (v) Comments or notes, e.g. lies in passage, calved in field.
(l)   Drying off dates and antibiotic used (if any).
(m) Dates culled with reasons. Give scores out of 10 for the reasons, e.g. 6 for infertility, 4 for low yield; destination; value.

All this information should be noted in the herdsman's diary as it happens and transferred on to the individual cards at least once a week. The herdsman might also have to record the dates of use of natural service (and, of course, the bull used) or the dates when the bull was put with a group of cows.

Cows are seldom weighed, and probably won't be until such a chore is automated. However, it is relatively simple during routine examinations to 'condition score' the cows at critical times during their annual cycle, and this provides valuable extra information.

An example of an individual card design is shown in fig. 7.1.

Rotary boards

If the card system is not very visual, or if it is kept away from the herdsman's office, then a strongly visual record reflecting the herd position now, and allowing for certain predictions for the future, is necessary.

There are several commercial makes of rotary (or rectangular) board available using large flat-headed pins or magnetic dice to reflect the cow's status at present. The future calving pattern can be seen at a glance as the pins for the cows diagnosed in calf are moved to the predicted calving date. The failure of the cows to show heat, be served or to conceive is highlighted by the colour of the face of the disc or its place on the board. The information is only transitory, little analysis is possible and cause and effect cannot be examined.

List of main events in calving date order

In addition to the card per cow kept for each animal, it is worth keeping a list of all cows on a chart on the wall (fig. 7.2). This list should be large enough to allow the following information to be shown in columns across the page:

Cow ID
Calving date
Lactation number
Bulling not served
Services
Pregnancy diagnosis
Culling date
Calving to conception intervals

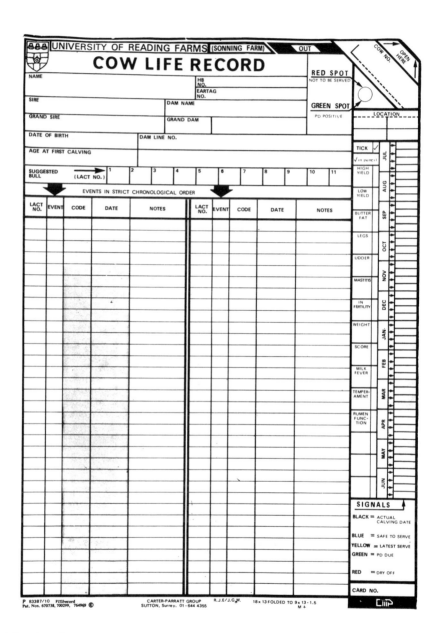

Fig. 7.1 Cow life record

Source: University of Reading Farms

| | | | | | UNIVERSITY OF READING FARMS | | | | (SONNING FARM) | | | |
| | | | | | | | | | | | | | |

EVENTS IN STRICT CHRONOLOGICAL ORDER

LACT NO.	EVENT	CODE	DATE	NOTES	LACT NO.	EVENT	CODE	DATE	NOTES

REMARKS :

P 83388/10   VISrecord
Pat. Nos. 670738, 700299, 764969 ©

CARTER-PARRATT GROUP
SUTTON, Surrey. 01-644-4355

R J E/J.G.M.

1 NAME OR NUMBER OF COW	2 CALVING DATE	3 SEX AND ID. OF CALF	4 CALVING COMPLICATIONS REQUIRING POST-NATAL VET CHECK E.G. RETAINED AFTERBIRTH, DIFFICULT CALVING, VAGINAL DISCHARGE, UTERINE OR VAGINAL PROLAPSE. 16-21 DAY TARGET DATE.	5 POST-NATAL VET CHECK. DATE DONE AND TREATMENT IF ANY	6 42 DAY POST-CALV-ING TARGET DATE, BY WHICH POST-NATAL COMPLETED AND FIRST HEAT RECORDED	7 IF NO RECORDED HEAT BY COL. 6 TARGET DATE, VET TREATMENT AND DATE DONE E.G. PROSTA-GLANDIN, KAMAR	8 RECORDED HEAT DATES		
							1	2	3
294	1/8/34	WK192 FR.F	Diff. Calving Toin	PNC-7/9	7/10 OK	KAMAR 7/10	12/10		

9 49 DAY POST-CALV-ING TARGET DATE AFTER WHICH SERVICE SHOULD BE DONE IF HEAT OBSERVED	10 SERVICE DATES BULL ID.					11 CALVING TO 1ST SERVICE INTERVAL IN DAYS	12 49 DAY TARGET DATE FOR VET P.D.	13 P.D. TO LAST SER-VICE	14 CALVING TO CONCEPT-ION IN DAYS	15 56 DAY PRE-CALV-ING TARGET DATE FOR DRYING OFF	16 DATE DUE TO CALVE	17 REMARKS, DISEASE INCIDENTS ETC.
	1	2	3	4	5							
19/10	26/10 BMR	11/11	Bull name			56	31/12	✓1/c	72	26/6	17/3	

Fig. 7.2 Example of sheet for herd records

Source: Dalgety Agriculture

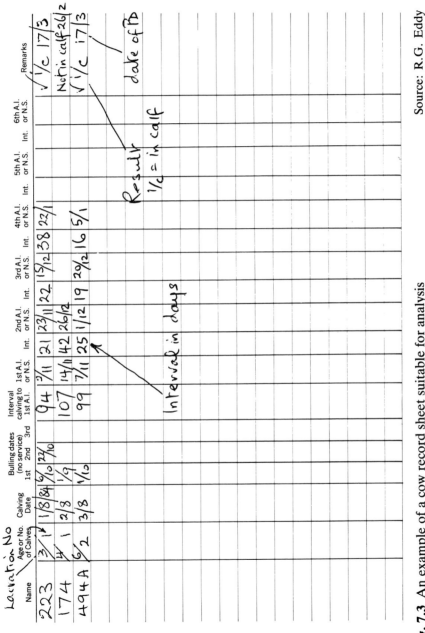

Fig. 7.3 An example of a cow record sheet suitable for analysis

Source: R.G. Eddy

The gaps between the events allow the insertion of calculated intervals between two consecutive events so that the crucial indices for the herd can be calculated (see fig. 7.3).

Heifer records

A record of the history of each cow as a heifer is also important and should include:

Dam's identity (and herd of origin)
Dam's lactation no.
M.A.F.F. ear tag no.
Identity of heifer as a calf (freeze brand no. later)
Date of birth
Weight at birth
Weight at weaning, first turnout, midsummer, housing, service, second turnout and calving (reasonable level of weight information plus condition score, height and comments at these times)
Dates served (and bull)
And/or date put with bull
Worm treatments
Any incidence of disease, injury, surgery, mastitis, etc., with any difficulty noted; calf mortality; date calved into herd – with remaining information being as per the cow records.

Milk recording

Any fertility recording system also needs good milk records and generally these should be taken at an evening and morning milking at least fortnightly throughout lactation. This information for each cow will allow a better understanding of fertility and health conditions. Firstly, rapid drops in yield will often indicate or predict some metabolic problem directly or indirectly caused by management of health; and secondly, absolute levels of yields will indicate if the cow is likely to have metabolic or even fertility problems. Very high yields (over 32 kg/day for heifers and over 40 kg for cows) may place the animal on the edge of a metabolic or fertility precipice calling for very steady, consistent management to help the animal through the period involved.

It is worth noting that *average* daily yields taken from cows 8–15 days after calving give a reasonable indication of the cow's peak yield (8–15 day *average* multiplied by 1.1 for first lactation heifers and 1.25 or 1.3 for cows). In addition, the peak yield in animals can be used to calculate the approximate lactation yield (305 days).

305
day
yield
{
1. Heifers – multiply peak by 240
2. 2nd lactation cows – multiply peak by 220
3. 3rd/4th lactation cows – multiply peak by 200
4. 5+ lactation cows – multiply peak by 190
}

A yield drop of greater than 10% in a week will usually indicate a management or health problem. Provided the records are accurate, the herdsman should inspect such cows carefully for mastitis, lameness, digestive problems, being bullied off the feed face, problems with the teeth and/or mouth, recent evidence of bulling, etc. The milk records should be added to the card and a graph kept of performance.

It is important to remember that what has been outlined here represents the basic information which should be available for any cow in any herd. Some methods of recording and storing it have been suggested, but there are many others and as long as they form a sound basis for monitoring performance and taking decisions then they will be adequate. The best record system is the one which suits the farmer who operates it!

Analysis (and computers)

Reproductive records, even when kept, are seldom analysed properly, and most farmers seem to consider their levels of performance to be about one third higher than they turn out to be on analysis and examination. Herds generally measure their fertility performance by calving indices; these are often taken from two-year-old information in the national milk records as the averages are produced from cows drying off between one September (the official end of the recording year) and the next (see fig. 7.4).

However, we now know what records to keep for each animal in the herd and what analysis to make. The limiting factor has often been the lack of time available to analyse even well kept records. There has been a gradual development recently of computer based systems to cope with fertility and health records. Some of these systems have *action lists* and also analysis of the fertility

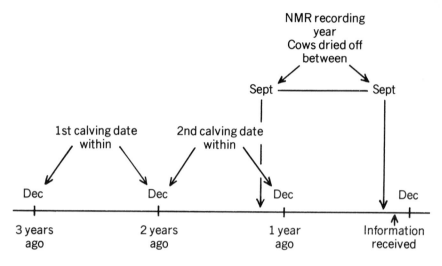

Fig. 7.4 Age of data involved in calculating calving intervals in National Milk Record Scheme

performance overall. But often, because of their 'main frame' centralised computer base, the turnround of information back to the farm is too long to be of much use and the data collection routine is often poor, leading to inaccurate information getting into the system and the production of worthless lists and reports. Where such difficulties occur there is considerable scope for the use of good hand worked schemes or, better still, of micro-computers to store, retrieve and analyse the information. In some cases these can be used in addition to the National Milk Recording Scheme which has its purpose in pedigree records and bull evaluation schemes through progeny testing.

Although microcomputers are now becoming cheaper, there are several areas of difficulty of a technical nature to be discussed before one can say that it is sensible to go ahead and install or use a system on the farm or in a veterinary practice. Firstly, computers need maintenance and support from computer engineers whose job it is to visit, service and repair such machines. Computers take time. Entering data into and operating a microcomputer takes about one to four hours per week per 100 cows depending on the details recorded, the skill of the operator, the way the program has been written and the type of computer. Thus many farmers find that buying a micro is one thing but operating it for the good of their dairy herd is another, and that it requires a disciplined and organised approach.

The computer will not itself manage the herd, since it will not make decisions itself but only provide information. The message it produces has to be carried by someone to the cow or cows in question and put into effect. The way to make use of this new technology has to be considered and planned right down to reaching the cow and helping her.

Choosing a computer

A computer should be chosen primarily on the strength of the programs or 'software'; the hardware should be secondary. Little else will eventually interest the user provided the program loads and works every time and there is enough disk capacity in the machine to cope with his programs and data. The system must, basically, be easy to operate and not break down.

The system should produce output quickly following rapid, easy data entry and printouts should be easy to read and understand with the minimum of codes and jargon. The quality of data should be maintained by plenty of 'built-in' checks.

Not all programs run on all computers. Rather than go out and buy a machine with a well known name and then hope the software that you like will run on it, first find the software that does what you want and then choose a micro that will run it – making sure that it has a large enough capacity to cope with the next two years' estimated growth.

Writing or having programs written is very time consuming and expensive. It is probable that a full dairy recording system has taken ten man years to write, validate, correct and finish off. It is better to buy commercially produced, tested software. Fortunately, most of the U.K. farming software has been produced by agricultural specialists and is of high quality.

In addition, it will become more and more sensible to consider systems with plenty of storage space using the hard disk medium as well as floppy disks. There are now portable hard disks which are relatively inexpensive and present a system that can cope with 10–20 million characters of information on a single hard disk.

Generally, a computer should be bought as a consumable item which will need replacing (or adding to) in two or three years.

Hence the need for transportable software. The computer should therefore have a specification of at least:

1. Local floppy disk storage of at least 750 kilobytes per disk for data.
2. A visual display unit of 24 × 80 characters.
3. A standard 'QWERTY' keyboard, numeric keypad, easy controls.
4. A printer of at least 150 characters per second print speed and able to print 132 characters per line.
5. A hard disk unit of at least 10 MB (megabytes), and ideally 20 MB, capacity to allow the operation of a range of operating systems like BOS/5, MS-DOS or UNIX, so that the full range of software can be operated.
6. Good servicing and support.
7. Wide range of other software available for it to cope with accounts, stock control, spread sheet calculation, etc.

Fertility recording and reporting using microcomputers

The available dairy software system should be carefully reviewed and the following questions considered:

(a) How easy is it to enter back data for each cow quickly, i.e. can summaries be entered, can only part of the data be entered?
(b) Can data be entered in any order for any cow?
(c) Can milk records be entered by day of recording?
(d) Can data (e.g. P.D.s) be entered in batches without un-necessary repetition?
(e) Can data be corrected quickly?
(f) Does the system help validate the data?
(g) Are the codes used extendable and changeable while still allowing some commonality for all users of the system?
(h) Does it allow any number of events e.g. 15 services?
(i) Does it allow the user to specify the codes?
(j) Does it allow new diseases, findings, diagnoses and treatments to be added?
(k) Does it allow any number of bulls to be used?
(l) Does it allow a revisit system to be set up for treated cows?
(m) Does it allow data to be stored indefinitely?
(n) Does it allow an unusual sequence of events to be entered?
(o) Most important of all, can it accommodate all the recordings you are accustomed to making or those you wish to make?

Entering back data

Once a system has been chosen and bought then it is important to enter the herd's back data. This entails going back with records to the last calving date and entering a reasonable amount of information for all the cows currently in the herd, such as:

Cow no.
Lactation no. at present
Calving date
Main events since calving
   Services (inc. date and bull)
   P.D. results
   Mastitis
   Drying off

If possible this information should also be entered for cows calving in the last twelve months that have since been culled. This gives the opportunity for the analysis to show the complete level of performance in the herd.

Some general considerations

A system needs to be devised to ensure that the records are kept regularly, completely and simply. This probably means the elimination of time-consuming alternative recording systems if possible. Thus the service chart may not be necessary and the individual cow card can be dispensed with *provided* the computer system can produce the alternative in the form of a printout quickly, regularly and better analysed or sorted.

The day-to-day book or diary used for the raw data (see fig. 7.5) needs to be small enough for the pocket, large enough not to lose, strong enough to stand a lot of use, and in a duplicate form so that, without carbon paper itself being necessary, a permanent set of original records remains in the book. The record is entered as fully as possible, using the codes suggested wherever possible but otherwise adding comments to allow an experienced coder to enter the information. A good system will not demand strictly chronological records, and this is helpful as some notes about calvings or mastitis get temporarily mislaid or forgotten and yet must be entered.

Herdsmen will keep records as often as the information is

Daisy Record Sheet №. 012223

		HERD No.	**456**

COW	CODE	DATE	EVENT
123	CALV	1/9/84	Calving Live FR. Heif
292	SERV	1/9/84	Served H.J.P.
114	MAST	2/9/84	Left Front. Mastitis.
445	PD+	2/9/84	in calf
19	VLD	2/9/84	Vulval Discharge Pessary See 1wk.
242	ONO	2/9/84	Oestrus Not Observed CLLO Estrumate

Fig. 7.5 Daisy Record Sheet – example of herdsman's notebook

removed and analysed. If accurately recorded information is wanted, it should be removed and analysed frequently.

It is useful analysis that stimulates further recording in a circular process, so that one element feeds the other. The analysis is the reward to the herdsman for keeping the records, and it is important to involve the herdsman directly when setting up any scheme. He will tend to resist any innovation that is handed down to him 'second hand' by the farmer as he will see it as a threat imposed by a controller rather than as a help to himself.

Analysis of information

After entering one or two seasons' information for the entire herd, some analysis is possible. A variety of information becomes available and each component can be compared with agreed targets. Areas where immediate improvement is possible are thus highlighted.

It should be emphasized, of course, that such analysis should be performed whether or not a computer is used. Computers only make it quicker and simpler and therefore more likely to happen.

The following is a list of facts about the herd which should be studied in order to assess performance:

No. of cows calved
No. of heifers calved
Age at calving (heifers)
Total calved
Calf mortality
No. served since calving
Percentage of cows served since calving
Calving to first service interval median (days)
No. of cows served at less than 40 days post calving
No. of cows conceiving (total)
Percent conceived of those served
Percent conceived of those calved
1st service pregnancy rate
Calving to conception interval median (days)
Interval from first service to conception median (days)
Percent inter-oestral intervals at less than 17 days
Interval from start of breeding season to conception
Heat detection rate (% returns at 16–28 day intervals)
Herd reproductive performance index (H.R.P.)
Cows with retained afterbirth (% of those calved)
Cows with endometritis (% of those calved)
Cows with oestrus not observed treatment (% of those calved)

Cows with failure to conceive (% of those calved)
Pregnancy diagnosis negative rate (%)
Cystic ovaries (% of cows calved)
Cows died (no. and % of cows calved)
Cows sold (no. and % of cows calved)
Total culling rate (no. and % of cows calved)

All this information should relate to the number of animals calving in the period in question and should also be broken down to show the performance not just by the year but by months of calving.

For a quick check of the current position in terms of fertility management, the following analysis is worth producing.

	Targets
No. of cows P.D.'d in period	
No. P.D. positive	95%
No. P.D. negative	5%
No. P.D. recheck required	0%
Interval to conception (days)	85
Serves per conception	1.6
Submission rate (first 21 days of breeding season)	90–95%
Heat detection rate (% return at 16–28 day intervals)	75%
Pregnancy rate for cows served and P.D.'d (1st, 2nd and 3rd services separately)	55%

Further fertility analysis

Where a problem is suspected or detected the foreman, herdsman or adviser should be able to go down another 'layer' to examine the evidence in detail. The reports available should allow the following to be carried out without difficulty, but only when required:

(a) *Pregnancy rate* by *date* and by
       bull used
       service number
       lactation number
       physical group of the cow at the time of service
       days post calving
       inseminator

heat signs used
fertility treatment preceding insemination
days of the week
no. of animals served on the same day (with and without any
fertility treatments such as prostaglandin preceding)
(b) *Heat detection rate and accuracy*
    by time of season (by day)
    by group animals were in at the time
    by person seeing the cow in heat
    by stage post partum of the cow in question
(c) *Culling rate*
    by reason sold or died
    by month of calving
    by lactation number
    by month of sale

Analysis of the effect of different fertility treatments on return to pregnancy should cover:

- Percentage served following treatment
- Conception rate after treatment
- Treatment to conception interval
- Culling or wastage rate following treatment

All these lists may look somewhat daunting and would not be tackled except with computerised systems. They mean that more definite answers are provided instead of vague impressions but, even so, it is the skill of the interpreter that will make or mar the process. Too small a supply of numbers of animals in any category often prevents firm conclusions from being drawn. The interpreter has, of course, to know what report and analysis to call for, how to read the printout, how to ensure that only relevant cows are included (e.g. by eliminating cows not yet P.D.'d from the analysis of conception rates) and, above all, how to break it to the herdsman or farmer that, for example, his heat detection is poor or his favourite bull is infertile. Above all, of course, he has to develop the skill to be early enough with his appraisal to be useful in the future. He has to know what to do next to improve things. He also has to be able to take a view on what will be the likely effect of not reacting quickly, i.e. what will this do to the calving pattern or culling rate, how much is it worth in lost profit, how much can we spend on putting it right, will what we try actually put it right, how do we ensure that what we try to do now is carried out efficiently by the herdsman in future?

For some, it may be helpful to have the analysed information as well as the targets displayed pictorially in graph, histogram or pie chart form. Another way of displaying the information is to use the CuSum (Cumulative Sum) and Cumulative Difference techniques. However, not everyone finds graphical information any easier to cope with than the written word, so they are not necessarily the panacea.

Action lists

The following reports are produced to allow routine operations on the farm to be kept up with.

Cows due to dry off this month. Should be listed showing their number, group, yield this lactation, and to show the cow's latest condition score. In addition, the list should highlight any cows that have had mastitis in the current lactation. Cows with three or more separate cases in the same quarter might be considered as potential culls.

Cows due to calve this month. Should be listed showing any difficulties that occurred at the last calving. These difficulties include dystokia, milk fever, mastitis, cases of twins and retained foetal membranes. All of these conditions are repeatable, if only at a low rate.

Cows calved 35 days and more ago and not seen in heat. Should be listed so that action can be taken by the herdsman or vet (tailpaint or Kamar applied on veterinary inspection). The cow may not necessarily be non-cyclic but may simply not have been seen in heat. Any difficulties around calving should be listed.

Cows calved at 70 days or more and not yet served. Should be listed. (This list is sensible only if the serving season has been in operation for 24 days at least.)

Cows served (and not returned to service or heat) about 35–55 days ago. These cows are those for pregnancy diagnosis by the veterinary surgeon. The list should show the date of the last service, the last bull used and total number of services. The last calving date should also be mentioned so that decisions can be made, if necessary, about culling; cows more than 190 days after calving may be considered candidates for culling.

Cows found not in calf by the veterinary surgeon and which have

not been served again for at least 21 days since that inspection.
Should be listed. Here the date of the P.D. and any treatments
should be printed.

A running list needs to be kept up of cows that have been
designated 'not to be served'. These cows are sometimes on the list
temporarily (usually while ill), or in the close season for serving
they are the cows that either straight after calving or after several
services the farmer decides should not be served again.

Lists also need to be kept up of the fertility status of the herd.
These should include:

Cows status categories
    unserved no. and percentage of herd
    served not pregnant
    pregnant and milking
    pregnant dry
    unknown
    barren or set barren
Cows pregnancy diagnosed (in last 4 weeks, 4 weeks before that
and since beginning of season)
    no. P.D.s
    no. P.D. in calf (%)
    no. P.D. negative (%)
    no. P.D. uncertain (%)
Days to conception for those in calf (days)
Services per conception for those in calf (days)
Pregnancy rate (last 4 weeks, previous 4 weeks, whole breeding
season)
    first service no. services (%)
    second service no. services (%)
    third and other service no. services (%)
    all services no. (%)

In addition it is important to put the right cows forward for day-
critical events so nothing gets forgotten. These lists should include
(by day of the week):

Cows for milk progesterone test (day 24 after service).
Cows for second prostaglandin injection (day 12 after the first).
Cows for timed A.I. (days 3 and 4 after second prostaglandin).
Cows to have PRID coil removed (e.g. eight days after
insertion).

Cows not yet served, not seen in heat and needing tail paint (more than 21 days after calving).

Finally, the lists should include the cows for the veterinary surgeon to see on his regular visit (see chapter 4). The cows should be shown in cow number order and should include cows for revisits following previous veterinary (or herdsman) treatments. It is important to provide such a check of their effectiveness. Often this is not done, and all the expense of the first efforts goes to waste. Some system must be devised to ensure that these follow-ups are carried out. With a card system, a coloured tag should be clipped to the card and left there until the cow is clear of the problem. Sometimes a cow is needed for a revisit for more than one reason and it will be worth having enough tags to show this clearly.

Regular veterinary appreciation of the state of the herd

It is important for a plan to have been drawn up for the season so that any failures can be highlighted quickly and effective action can be taken on the farm. While many of the lists mentioned above are helpful in sorting out which cow to extract i.e. the tactics needed, the system will not necessarily highlight whether the strategy is working. Summarised analysis is needed to see if enough cows are being got back in calf each week or month to keep the calving pattern stable and the culling rate down. If these levels are disappointing (a degree of slip may be allowable but the interference levels should be clearly defined) then action must be taken. To take action one needs to know which actions can be chosen and their likely effectiveness. As the serving season runs out, more and more of the possible aids and techniques should be brought into operation to complete the season on time with the minimum of wastage.

Minimising errors in the system

Several types of errors can arise in any recording system:

- Errors of omission.
- Errors of misidentification of the animal at observation or at recording.
- Errors of transcription or at entry.

There is often no way of discovering any error until the data is analysed, either by the next stage of a hand system when a record is entered on to a card and an oddity noticed (e.g. this cow is already in calf and a P.D. result exists for her – perhaps it's not 371 but 317) or by well written validation routines which are incorporated in good computer systems.

A roving visiting recorder to check the information 'by eye' when collecting it is a help with hand-kept systems but the effectiveness of such a person is minimal compared with sophisticated computerised schemes that list the uncorrected errors. These errors need correcting before the reports are run.

If the reports contain errors, confidence in them will be minimal and the herdsman will use other methods to cope. The error rate must be less than one per cent. For crucial factors (such as cows not yet served, due to dry off or to calve, to treat with prostaglandin) it really has to be zero per cent as mistakes here can mean dead calves, lost embryos and the loss of a great deal of money. The lines of communication between the herdsman writing down the raw record and the computer operator have to be open and very effective.

Errors also occur at the computer, however. For example:

- Data misread by the operator.
- Incorrect keys punched.
- Data entered wrongly and not corrected.

In conclusion, then, records and their analysis are the basis of sound and effective management. Effective recording demands a conscientious approach and good co-operation among all farm staff; and regular analysis is what makes it all worthwhile. Remember, whether using a computer or a so-called 'manual' system, the analysis is only as good as the information which has produced it!

8 Is it all worth it? A case study

Reported studies of planned animal health programmes have generally shown that calving to conception intervals in herds visited can be reduced by 15–25 days, down to the target 85–100 days, depending on the starting point.

While many of the earlier studies concentrated on uterine involution and treatment for various conditions in the breeding tract, later studies have emphasised the importance of heat detection and the recording of key events. Suboestrus and anoestrus are today considered to be problems of the herdsman rather than of the cow, but the important feature is to diagnose the 'condition' of poor fertility management and identify the offending component.

Essential elements

The facilities that must be provided for a fertility management scheme to work properly are:

1. Proper cattle handling facilities to permit rectal palpations.
2. Complete and legible cow identification (freeze branding and ear tagging).
3. A recording scheme operating for all key events and health and fertility events.
4. A monitoring system, analysing performance so that the veterinarian or farmer and herdsman can recognise the current level of performance and whether targets are being achieved.

Above and beyond these requirements, the key to success lies in the attitudes and skills of the people involved.

The herdsman

In one study it was found that the crucial element affecting the level of fertility performance and the operation of the control programme was the herdsman's skill, judged as follows:

- Attitude to fertility management.
- Understanding of the bovine oestrous cycle, use of prostaglandins and other treatments.
- Skills of organisation and application of management techniques.
- Susceptibility to advice from the farmer, veterinary surgeon or others.
- Relationship with the farmer and veterinary surgeon.

The veterinary surgeon

Even very highly skilled, experienced veterinary surgeons may be unable to overcome the limitations of a poor herdsman. In cases where the herdsman is the limiting factor it can be possible to get round some of the problems by the use of the skills of the veterinarian and the products at his disposal. One study showed that the judicious use of prostaglandins, the milk progesterone test and Kamar Heat Mount Detectors did reduce the calving interval by 20 days and the culling rate from 30% to 22% even where the herdsman had refused to co-operate (particularly over late-night observation of heat).

Regular review meetings are essential with fertility control schemes as the problem is multi-factorial. All those involved should be present and the latest performance discussed so that everyone can appreciate whether tactics to improve fertility are working. Someone needs to have analysed the latest set of records fully so that all the evidence is to hand, in order to plan which future techniques can be tried before the next review meeting. Such meetings should take place at regular intervals, and during difficult times can profitably be carried out monthly.

In any scheme it may be worth grading herds in terms of a fertility index (see table 8.1). One used by several workers is:

$$\text{FI (fertility index)} = \frac{\text{Pregnancy Rate to 1st Service}}{\text{Services per conception}} - (\text{calving to conception} - 125) - (\text{culling rate} - 25)$$

The target set by a M.A.F.F. committee for improvement in herd fertility was 12 days off the calving to conception interval without affecting the culling rate. This was practically achieved in

Table 8.1 Table showing grades according to fertility index (F.I.)

Grades	Score	Requirements to achieve grade
High standard	86+	Calving to conception less than 90 days Pregnancy rate first service more than 60% Culling rate less than 20%
Good	71–86	Slightly poorer in all three components or very poor in one
Average	60–71	Calving to conception 90–95 days Overall pregnancy rate 50% Culling rate 18–25%
Poor	< 60	Poor pregnancy rate 34–52% Culling rate 17–33% Intervals to first service longer than 70 days on average

Source: R.J. Esslemont and P.R. Ellis, University of Reading. Study No 21. 22 herds over three seasons.

Table 8.2 Improvements in fertility performance in 28 herds over two years

	Calving to conception	% cows sold or died
Base year	102.5	12.3
Year 1	96.2	18.6
Year 2	91.6	16.9

Source: M.A.F.F.

the national study of 28 herds which was set up to try to improve herd performance (table 8.2).

In a major study of a fertility control programme over several years in 2000 cows, a practising veterinarian in Somerset saw the calving to conception interval reduced by 15 days (from 105 down to 90). The proportion of the 'herd' conceiving (of those calved) remained unchanged at 84% and the overall pregnancy rate varied between 46% and 52%. The majority of the improvement in calving to conception resulted from an earlier first service, and only a small proportion from a shorter interval between first service and conception. It was found in this study that oestrus detection rates (table 8.3) could be improved for first services but not for returns to heat.

Table 8.3 Aggregated performance of 10 herds over five years on a fertility control programme

	Year				
	1	2	3	4	5
No. calved	1127	1492	1691	1801	1794
No. served	1057	1395	1558	1652	1651
No. conceived	965	1238	1423	1447	1508
% served of calved	93.8	93.5	92.1	91.7	92.1
Calving – 1st service	76.3	75.0	70.2	66.6	64.0
Calving to conception	105.5	103.0	93.5	93.4	90.1
% conceiving of served	90.4	88.7	91.3	87.6	91.3
% conceiving of calved	84.8	83.0	84.2	80.3	84.0
First service preg. rate	51.9	53.0	55.0	50.6	52.5
All served preg. rate	46.5	50.0	52.0	47.0	50.2
Interval					
1st service conception	29.2	28.0	23.3	26.8	26.1
Herd reproductive performance (H.R.P.)*	40.7	43.7	63.6	58.2	69.7

* Herd reproductive performance from R. G. Eddy, Fellowship thesis (F.R.C.V.S.)
H.R.P. = (125–calv.–conc.) + (preg. rate – 45) + (80 – (c – 1st S)) + (% conc. of served–80)
Commonly accepted target: H.R.P. = 40 + 20 + 20 + 15 = 95

Park Farm case study

Introduction

Park Farm extends to 552 hectares of free draining chalk loam and comprises two dairy herds, dairy replacements and a large arable acreage.

Grassland and forage	250 hectares
Arable	193 hectares
Woodland	100 hectares

Both dairy units, Park and Manor, are two man units of 200 cows each; Park has cubicles and straw yards and a 10 × 10 herringbone parlour and Manor has a 14 × 14 herringbone with A.C.R.s. The dairy heifer rearing unit copes with 100 heifers each year and 80 of them are used as replacements in the two herds. All bull calves and the remaining heifers are sold at one week old.

There is a block of 80 hectares of grassland close to each dairy unit in the farm of long term leys which are set stocked. The young stock graze ley breaks within the arable rotation.

Management policy

The present structure outlined above stemmed from three herds, one of which was in a cowshed and was disbanded in 1977 following a change in management. At the same time it was decided to alter the youngstock rearing policy to two year calving instead of three. To do this the group of heifers aged from 12–24 months were all sold. The older group (25–36 months) were calved into the two enlarged herds while the younger group (0–12 months) were put on to a management routine to produce 510 kg heifers to calve at two years. This move released 40 hectares for cereal production and reduced by one third the number of young stock carried on the farm.

In the first year a plan was drawn up to tackle the fertility management. The aims were to:

• *Move the calving pattern* from a well spread autumn/winter policy (July to May!) to a more closely defined and profitable period (September to December).
• *Reduce average age at calving* from 33 months to 25 months.
• *Reduce calving interval* to 365 days from 395 days.
• *Maintain culling rate* at around 20%.

The methods used to achieve these targets were:

1. Serve cows from 15 November to 1 June, reducing the length of the season by two weeks each year. Once sufficient cows were put in calf to produce replacements, a Hereford bull was used to 'sweep up'. Cows not in calf by 1 June were culled.
2. Serve all cows after 42 days post partum (provided they were in the serving season). All cows fit for service on the first day of the breeding season were noted and any cows seen in heat inseminated

on detection. The cows not seen in heat by day 7 were examined by the veterinary surgeon and given an injection of prostaglandin. These animals were then served on detection of heat. Prostaglandins were used subsequently on remaining cows in 'fortnightly' batches as the cows reached 42 days post calving. Returns to service were detected and the cows inseminated normally.

During the serving and pregnancy diagnosis season the veterinary surgeon visited fortnightly to attend to cows. The cow records were processed every two weeks by the Daisy bureau service from the University of Reading. The turnround of the records was kept to 48 hours from farm to farm. The action lists were used to identify the cows suitable for injection with prostaglandin and to highlight cows not yet in calf for one reason or another.

Results

The proportion of the herd calving by the end of the calendar year improved from 58% to 85%, while the culling rate varied from 17–21% over a four-year period (table 8.4).

Table 8.4 Calving pattern of the herd over the years of the study

Season of calving	Total no. calving	No. calved by 31 Dec.	% calved by 31 Dec.	Culling rate %
Year before beginning	191	111	58	17
1st year	198	133	67	21
2nd year	190	130	68	21
3rd year	195	146	75	20
4th year	195	166	85	21

Apart from the first service pregnancy rate (42%), the indices were very good. Calving to conception was 93 days and the proportion conceiving again was 86% (fig. 8.1 and table 8.5).

The age of the heifers calving into the herd declined from 33 months to 25 months, and because the heifers themselves were synchronised and served with dairy semen the proportion of heifers born by 30 November increased from 60% to 95% (table 8.6). The pregnancy rate to the synchronised service in the heifers never fell

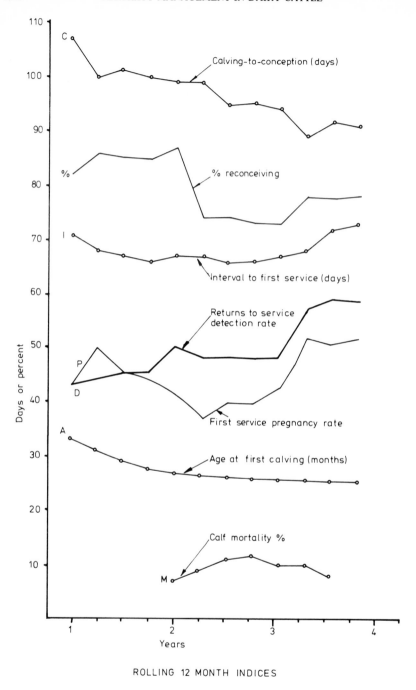

Fig. 8.1 Performance in Park Farm case study over four years

Table 8.5 Fertility performance fourth season and first season

	First	Fourth	Target
No. calved	200	195	
No. served (%)	180 (90%)	191 (97%)	95
Calving to first service (days)	75	65	65
First service pregnancy rate (%)	48	42	60
No. conceiving	160	166	
Per cent conceiving of served	88	86	95
Per cent conceiving of calved	80	85	90
Calving to conception interval (days)	109	93	85
Serves per conception	1.7	1.9	1.6

Table 8.6 Heifer performance over four seasons

Season	No. of heifers calving in to herd	% of herd	Age at first calving (months)
First	31	16	33.2
Second	40	21	31.9
Third	40	20	25.0
Fourth	40	20	25.5

Season	Total calvings	No. Friesian female calves retained	Proportion born by 30 Nov. %
First	198	50	60
Second	190	55	96
Third	195	54	96
Fourth	195	50	95

below 60% (table 8.7). In the cows the pregnancy rate to the single injection routine was the same as that herd's normal pregnancy rate (45%), with the peak of service activity occurring on the third and fourth days after treatment (table 8.8).

Table 8.7 Use of Estrumate in heifers

	No. of heifers 2 + 2 regime	Conception rate to 2 + 2 regime
First year	43	76
Second year	42	60
Third year	48	65

Table 8.8 Response to injections in one year (no. treated 150)

	No. A.I.'s on days after injection									Total A.I.	No. to 2nd P.G.
	0	1	2	3	4	5	6	7	>7	A.I.	2nd P.G.
No. served	7	1	6	48	25	15	4	1	2	109	41
No. conceiving	3	1	3	25	8	7	1	1	0	49	17
Percentage	43	100	50	52	31	47	25	100	—	45	41

The financial performance (table 8.9) improved in 1984 terms by £445 per hectare (£206 per cow) which led to improved margins for the herd of about £40000. This was the result of a better calving pattern, leading to a more simplified and effective management and feeding system for the herd, with more milk being produced per cow and with concentrates used more efficiently.

Table 8.9 The effect of improved fertility and other management on margin over purchased feed and forage in fourth year of study with year before study

	Year 4 of study	Year before study	Difference	
No. of cows in herd	196	209		
Stocking rate cows/ha	2.0	1.91		
Milk				
Litres/cow	6005	4565	+1440	
Pence/litre	15	15		
£/cow	901	685	+ 216	
Concentrates				
£/tonne	160	160		
Tonnes/cow	1.84	1.72	﹨ +	0.12
Kg/litre	0.31	0.38	−	0.07
£/cow	294	275	+	19
Margin over purchased feed				
£/cow	607	410	+ 197	
Nitrogen				
N use kg/ha	207	240	−	33
£/ha	82	95	−	13
Margin over feed and forage				
£/cow	566	360	+ 206	
£/ha	1132	687	+ 445	

Using constant prices and updated to 1984

Terms and definitions

Abortion The production of one or more calves between 152 and 270 days after an effective service which are either born dead or survive for less than 24 hours.

Assumed pregnancy rate The number of cows or heifers served within a defined period not observed to return to oestrus before a specified date expressed as a proportion of the total number of services given over that period. The defined period should finish at least 60 days before the end of the period for which data is available.

Average herd size The average of the number of cows in the herd as counted on twelve or more approximately equally spaced occasions during a year.

Bull An entire male aged 180 days or more.

Bull calf An entire male of less than 180 days of age.

Calving The birth of one or more calves more than 270 days after an effective service.

Calving index The mean calving intervals of all the cows in a herd at a defined point in time, calculated retrospectively from their most recent calving at that time.

Calving interval The interval, in days, for an individual cow from one calving to the next.

Calving rate The number of services given to a defined group of cows or heifers or over a specified period which result in a calving, expressed as a percentage of the total number of services.

Calving to conception interval The interval, in days, from calving to the subsequent effective service of a cow.

Calving to first service interval The number of days from calving to the first subsequent service of a cow.

Cow A female after the start of her first lactation.

Cow not to be served A cow which it is not intended to serve again and is destined to be culled.

Cull cow A live cow transferred out of the dairy herd irrespective of the purpose to which she is put subsequently. Cull cows may be separated into two groups: those culled before service and those culled after service.

Culling rate The number of cows calving in a defined period (usually twelve months) which are transferred live out of the herd before starting another lactation, expressed as a percentage of the total number of cows calving in the period.

Dairy herd One or more cows milked, managed and recorded as a single unit.

Date of conception The date of the effective service.

Date of service The date of the first natural mating or artificial insemination during a period of continuous oestrus.

Earliest service date The date on or after which a cow observed in oestrus would be served as a matter of policy.

Effective service A service which results in pregnancy.

Embryo The product of conception (the conceptus) from the date of conception to day 42 of pregnancy.

Embryo loss The loss of a conceptus during the first 42 days of pregnancy.

Foetal loss The loss of a foetus between 43 and 151 days of pregnancy.

Foetus The developing calf from 43 days of pregnancy to birth.

Heifer A female aged 180 days or more which has not started her first lactation.

Heifer calf A female of less than 180 days of age.

Herd size The number of cows present in the herd on a given date.

In-calf heifer A heifer which is confirmed in calf.

Inter-service intervals The number of days from one service of a cow to the next in the same lactation.

Maiden heifer A heifer which has not been served.

Mean calving to conception interval The average of the individual intervals of a group of cows calving over a defined period.

Mean calving to first service interval The average of the individual intervals for any defined group of cows.

Non-return rate to first insemination The number of first inseminations given to a defined group of cows or heifers or over a specified period of time which is not followed by a repeat insemination within a prescribed period, expressed as a percentage of the total number of first inseminations during the period.

Oestrus cycle The regular recurrence of oestrus together with related changes in the genital organs and the reproductive hormones.

Oestrus cycle length The number of days from the start of one oestrus to the start of the next. The day of the start of oestrus is counted as day 0.

Oestrus The physiological state in which a cow or heifer will stand voluntarily to be mounted.

Overall pregnancy rate The number of services given to a defined group of cows or heifers or over a specified period which result in a diagnosed pregnancy not less than 42 days after service, expressed as a percentage of the total number of services. Services to cull cows should be included. The method of pregnancy diagnosis must be specified.

Percentage culled and died The number of cows in a herd calving in a defined period (usually twelve months) which are culled or die before starting another lactation, expressed as a percentage of the total number of cows calving in the period.

Percentage pregnant of cows served The number of cows calving in a defined period (usually twelve months) which are diagnosed pregnant not less than 42 days after an effective service, expressed as a percentage of the number of cows served. The defined period should be the same as that used to calculate the mean calving to conception interval.

Predicted calving interval Calving to conception interval + 280 days (mean gestation length), calculated forward from the calving date. The mean predicted calving interval is the predicted calving index.

Pregnancy rate to first service The number of first services given over a stated period or to a defined group of cows or heifers which result in a diagnosed pregnancy not less than 42 days after service, expressed as a percentage of the number of first services in the period. The method of pregnancy diagnosis must be specified.

Premature calving The production of one or more calves between 152 and 270 days after an effective service, at least one of which survives for 24 hours or more.

Replacement rate The number of cows or heifers required to replace cows which have left the herd during a defined period (usually twelve months), expressed as a percentage of the average herd size during the same period.

Reproductive efficiency The number of cows becoming pregnant in a 21 day period expressed as a percentage of the number of cows eligible for service at the start of the period.

Served heifer A heifer which has been served or which has received a transferred embryo but which has not been confirmed in calf.

Service One or more natural matings or artificial inseminations during a period of continuous oestrus.

Services per pregnancy The total number of services given to a group of cows or over a defined period, divided by the number of services which result in a diagnosed pregnancy not less than 42 days after service. Services to cull cows should be included.

Stillborn A calf born dead or found dead after an unobserved calving.

Submission rate The number of cows or heifers served within a 21 day period, expressed as a percentage of the number of cows or heifers at or beyond their earliest service date at the start of the 21 day period.

Explanatory notes on terms and definitions

Abortion, premature calving, calving Where an abortion, premature calving or calving follows embryo transfer, the date of the effective service should be taken to be the date of the first observation of oestrus of the receipt immediately preceding the transfer.

Cull cow Although it is realised that live cows are on occasion transferred to other herds for further breeding, these are included in order to clarify the keeping and interpretation of herd records.

Dairy herd Units containing only one cow are included in the definition of a herd in order that they can be accounted for in calculations of mean herd size on a national or regional basis.

Embryo The term used in this definition refers to all the tissues which are the products of conception and therefore includes foetal membranes.

Foetal loss/abortion The loss of a pregnancy before 152 days will not initiate a new fertility record. Cows losing a pregnancy on or after 152 days will start a new fertility record from the date of the abortion as if they had calved normally.

The voiding of a calf at any time before 271 days constitutes an abortion for the purposes of the official brucellosis control schemes and must be reported to the local Divisional Veterinary Officer.

Heifer The definition differs from that in colloquial usage where the term continues to be used throughout the first lactation and up to the second calving. Excluding cows in the first lactation avoids confusion when dealing with the total number of animals in milk and is in agreement with the definition used by major recording organisations. When first lactation animals need to be identified the terms 'first calver' or 'first calved cow' should be used.

Inter-service intervals Inter service intervals should be allocated to one of the following groups: 2–17 days; 18–24 days; 25–35 days; 36–48 days; and 49 or more days. The group of cows or the period over which the services were given must be specified.

Non-return rate to insemination Non-return rates to first insemination are used by A.I. centres to monitor the fertility of bulls and the performance of inseminators. They are normally assessed at 30–60 days or at 49 days after service. The results are usually some 20 percentage points higher than calving rates.

Pregnancy rate to first service The National Group on Reproductive Definitions in U.S.A. recommended that the embryo should be defined as the product of conception up to day 42. Although pregnancy diagnosis can be carried out before this time the results obtained do not give an accurate prediction of calving because of embryo loss.

Pregnancy rates and services per conception Where pregnancy diagnosis is not undertaken evidence of pregnancy such as subsequent calving or abortion is taken in order to calculate these indices retrospectively.

Service More than one service at a single oestrus cannot result in more than one pregnancy per cow. Multiple services at a single oestrus must therefore be considered as one service for the purposes of calculating pregnancy rates and services per pregnancy.

Source: H.M.S.O. Definition of Terms used in Fertility Measurement in Cattle

Index